Numbers & Math

Pre-K Teacher's Guide

8001 MacArthur Blvd
Cabin John, MD 20818
301.263.2700
getsetforschool.com

Authors: Jan Z. Olsen, OTR, and Emily F. Knapton, M. Ed., OTR/L
Illustrators: Jan Z. Olsen, OTR, and Julie Koborg
Curriculum Designers: Carolyn Satoh, Kate McGill, Cheryl Lundy Warfield, Suzanne Belahmira, Allessandra Bradley-Burns, Mónica Palacio, Adina Rosenthal, Robert Walnock
Graphic Designers: Jodi Dudek, Nichole Monaghan

The Get Set for School® program and teacher's guides are intuitive and packed with resources and information. Nevertheless, we are constantly developing new ideas and content that make the program easier to teach and to learn.

To make this information available to you, we created a password-protected section of our website exclusively for users of this teachers' guide. Here you'll find new tips, in-depth information about topics described in this guide, other instructional resources, and material you can share with students, parents, and other educators.

Just go to **getsetforschool.com/click** and enter your passcode: **TGNM11**

Enjoy the online resources, and send us any input that you think would be helpful to others: janolsen@getsetforschool.com

Dear Educator,

Welcome to the Get Set for School® Numbers & Math program. It's a program that's just right for Pre-K children. You may know us from our original Get Set for School Readiness & Writing program. If you do, your children are building letters with Wood Pieces, learning how to hold crayons, writing capitals and numbers, and singing "The Ant, the Bug and the Bee." Of course, they know and love Mat Man®.

Whether you are a long-time fan or brand new to Get Set for School, we know you'll enjoy this Pre-K Numbers & Math program. It's child-friendly and developmentally based. This is also a program that will fit into your day and your way of teaching. You'll find that the materials and activities in this program will help you stimulate and sustain learning.

This Numbers & Math program will suit all of your children and grow with them. You probably know that Get Set for School was started by occupational therapists who know and understand the readiness skills needed for later school success. We have high expectations, but we make learning accessible from the very first day. Now we have got math and early childhood experts working with us. Together we designed products, hands-on activities, and teaching strategies to boost number and math skills in the same active, child-friendly way.

This program respects children's need for self-directed play as it prepares them responsibly for kindergarten. This guide is organized by Pre-K teaching domains and learning skills/benchmarks so that you can have confidence your children are well prepared to meet kindergarten expectations. At the same time, we treasure children's natural curiosity and creativity. We respect what children already know and how much they learn independently. You'll find many suggestions for keeping their learning active and joyful.

We hope you have a wonderful year—a year in which learning and teaching is truly a delight.

All the best,

Emily F. Knapton   *Jan Z. Olsen*

Emily F. Knapton and Jan Z. Olsen

Table of Contents

Getting to Know Get Set for School®

Using Your *Numbers & Math* Pre-K Teacher's Guide

Pre-K Classroom and Children

1 - Number & Operations
I Know How Many

2 - Geometry
Places and Shapes

3 - Patterns & Algebra
See It, Repeat It

4 - Measurement & Time
Short & Tall, Big & Small

5 - Data Representation & Probability
We Can Show What We Know

Resources

Getting to Know Get Set For School®

Get Set for School Pre-K Philosophies and Principles

We took our years of experience and the most relevant research about how children learn best to develop this award-winning curriculum for school readiness.

Different and Better

We understand that preschoolers learn through movement and participation. They need explicit modeled instruction. They also need playful learning opportunities to explore and internalize new ideas. Our playful approach is at the heart of our success because young children are not ready to sit still and focus for long periods of time. They learn best when they move, manipulate objects, build, sing, draw, and participate in dramatic play. Preschoolers also need instruction tailored to their different styles of learning. You need tools to meet these needs. Our unique Pre-K programs make teaching easy and rewarding for you with:

- A research-based approach that addresses different learning styles
- Developmental progression that builds on what children already know
- Friendly voice that connects with children
- Multisensory lessons that break difficult concepts into simple tasks
- Hands-on materials that make children want to learn

What and How We Teach

Pre-K is a time of rapid growth and development. It is a time of preparation for kindergarten and for future success in school and in life. To achieve, children need to be **eager**, **able**, and **social** in their own learning.

Eager – Children learn naturally through everyday experiences with people, places, and things. They are born imitators and scientists who thrive on active hands-on interaction with the physical world. They learn through play and through physical and sensory experiences. The materials throughout the learning space affect how children feel, what they do, and how they learn. Our products (music and manipulatives) invite discovery and independent thinking.

Able – We deliberately build familiarity and competency with the music that we play, the words that we use, and the way in which we use materials and teach lessons. Our materials and lessons are flexible so that you can teach in a developmental sequence from simpler to more complex.

Social – Pre-K is social and challenging. We use music and hands-on materials to encourage inclusive participation and the development of social, physical, language, math, and early readiness skills. We read, draw, sing, build, and dance with children, enticing them to join us on a learning adventure. Our materials also encourage family involvement to continue learning at home. Many activities have take-home components to encourage children to form connections between learning in school and home situations.

A Developmental Curriculum

Pre-K children will enter your classroom with different and continually evolving abilities. They will not all be developmentally ready to learn the same skills at the same time.

To meet the needs of Pre-K children, a curriculum must be accessible at all points within this wide spectrum of needs and skills. It should invite participation, build a base of understanding, and challenge children's thinking. You need support to meet all your students where they are as they develop throughout the year.

Get Set for School teaches in developmental order by starting at a level that does not assume prior knowledge or competency. We enable children to excel by respecting their present level of development and building from there.

There are certain skills that children need to learn explicitly. We teach them how to read, write, and count. We help them to recognize, name, and write letters and numbers. We teach them how to listen for and say sounds. We also teach them important social skills.

Supporting Parents and Teachers

Parents are a child's first and most important teachers. Our materials encourage family involvement to continue learning activities at home. Our lessons sometimes ask families to lend items from home to personalize activities. Many of our activities have take-home components to encourage children to form connections between home and school.

Some of our materials are also designed to be taken home and shared with families as children practice skills learned in school, show their families what they are learning, and ask for their help and participation.

Get Set for School makes a seamless transition between home and school. We recognize that preschool is often a child's first experience away from home. Our activities encourage children to share family experiences and see how those become part of learning in school. Our curriculum also acknowledges and celebrates cultural differences.

Get Set for School® – Three Core Learning Areas

Get Set for School is a curriculum that prepares young learners for school with three complete programs: Readiness & Writing, Language & Literacy, Numbers & Math. These programs complement and expand your existing Pre-K program. You engage children with the following strategies:

- Creative lessons that enable children of different abilities to achieve
- Child-friendly language and activities
- Developmentally based teaching that works at every level
- Hands-on approach that promotes active participation

Get Set for School is recognized as the best children's curriculum and children love it. Their teachers, parents, and occupational therapists have asked for more. Now we have three core learning areas.

Readiness & Writing

This program is the core of Get Set for School. The handwriting component is based on more than 25 years of success with Handwriting Without Tears®. Writing requires many skills that are essential for school: physical, language, cognitive, social, and perceptual. Our Readiness & Writing program uses music, movement, and multisensory manipulatives to teach all the core readiness skills including crayon grip, letter and number recognition, number and capital letter formation, and body awareness.

Language & Literacy

This program is a wonderful complement to our Get Set for School Readiness & Writing program. We use dramatic play, singing, finger plays, manipulatives, and movement to teach children to rhyme, clap syllables, make and break compound words, and identify sounds. We expose children to rich literature (see Book Connections on page 172) to foster a love of reading, build vocabulary, and learn how books work. They learn facts from informational text. Children will learn to use new words and develop oral language skills by listening, retelling, and narrating stories. They learn that there is meaning in the words they say as they watch teachers write what they say.

Numbers & Math

This program is a natural extension of our Get Set for School readiness program and helps children build number sense right from the start. We use manipulatives, music, and rhymes to teach counting, comparisons, spatial awareness, patterning, sequencing, matching, sorting, problem solving, and even Pre-K geometry skills. Lessons give children time to play with real objects and test their ideas so that math becomes real and meaningful. Children also develop oral language skills that help them learn about and express math concepts.

Using Your Teacher's Guide

Curriculum Organization: Domains and Learning Skills/Benchmarks

Do you think of yourself as a math person, or does the word "algebra" give you the shivers? Take heart. We've compiled a friendly collection of resources and activities. We've got plain talk, simple explanations, and fun activities. We're also covering the essential math benchmarks. You and your children will delight in doing math together.

Math instruction is vital in the early years of a child's education. Numbers should be an integral and joyful part of children's classroom experiences. As you actively engage with the children, you build on and elaborate children's mathematical ideas. The Numbers and Math program provides activities that help children acquire critical early math skills. Activities are organized into five domains—groups of related learning skills. These learning skills are also called benchmarks. This guide includes two versions of benchmarks. One set is what we refer to as friendly titles. These are found in the table of contents, the first page of each domain, and in the skill descriptions on the left of each activity page. The second version is written in more formal academic language. These are the benchmarks you will see in the sections entitled, "Look What We're Learning." A complete list of these benchmarks can be found on pages 160–165 of this teacher's guide. The learning skills within each domain are organized from the easiest to learn to the more challenging. Activities and variations are provided for each skill. Children receive the support they need to learn and reinforce skills, make connections, and play independently.

Through spontaneous play and engaging activities, children develop early math skills in these five domains:
1. Number & Operations
2. Geometry
3. Patterns & Algebra
4. Measurement & Time
5. Data Representation & Probability

You may not be an expert in these math domains, but as you read this teacher's guide and work with our activities, you will become more confident in your knowledge. Here are some plain talk explanations on each domain.

Number & Operations

Pre-K children should develop number sense, investigate relationships among numbers, and explore the properties of numbers. Children know numbers in a practical way long before they do math activities. They know that they have one mouth and two hands before they recognize 1 and 2. They know that they want more, even before they know the word "more." What we teach children in Numbers & Operations are words and symbols for what they already know, while expanding their basic ideas about numbers to a solid understanding of quantities.

Geometry

Geometry is the study of shapes and space. When children play on the playground, they begin to learn words to tell where they are (e.g., on the ladder, under the slide). We want children to build their vocabularies with position words through songs, games, and activities. They also play, build, and explore with shapes. In Pre-K, children can move beyond simple identification of shapes to understand each shape's characteristics. Geometry and Number & Operations are foundational to future math learning, so Pre-K classrooms should spend a large amount of time on these domains.

Patterns & Algebra

Algebra is an area of math that uses symbols, letters, and patterns to solve problems. Children love to notice and make patterns. Seeing and extending patterns help children build observation, thinking, and problem-solving skills. Simple repetitive patterns and even simple growing patterns can be explored with young children. Pattern activities build the foundation for understanding more complex mathematical patterns in the future.

Measurement & Time

Measurement is determining the size or amount of something. It has direct application to everyday life. Young children acquire these skills through measuring objects themselves. Children can experience measuring by making direct comparisons between objects, comparing objects using nonstandard units, such as paper clips or straws, and comparing objects using standard units. They learn to measure with nonstandard units first to prepare them to work with standard units in the future. When Pre-K children learn about time, they think about general times of day and what happens in their lives at those times.

Data Representation & Probability

Data representation activities help children organize information (answers to questions) in a visual way. They are a good way to connect questions in children's real worlds with numbers. Pictographs can be created in response to almost any Pre-K question, such as favorite ice cream or number of pets. Once a graph is made, it is important to discuss the information in the graph briefly. Repeated exposure to ideas and data in pictographs improves children's understanding of comparison questions. Probability helps us answer questions about our world with regard to the likelihood of future events. It helps children make sense of their day and world.

Problem Solving

Problem solving is an important part of math. Children learn many math skills, yet without strategies, they won't know how to approach problems. Problem-solving strategies that work when children are four will still work when they're fourteen. While some strategies come naturally to children, others may need to be demonstrated and explained. When children get comfortable using problem-solving strategies, they can apply them repeatedly in math (and other real-life) situations.

Problem-solving strategies are embedded in activities throughout this program. They can be applied in any order, and more than one can be used to solve a problem. Each time a problem-solving strategy is used, it is highlighted and named in the Look What We're Learning section.

The problem-solving strategies introduced in *Numbers & Math Pre-K Teacher's Guide*:

Act It Out

Movement and speech can help children create the problem physically and find the solution. For example, children can act out addition by forming two groups and then having those groups join together. Everyone can see and count the separate groups and then the combined group. The process of physically combining groups can be powerful in helping children understand.

Use Objects

Much like acting out a problem, manipulatives can be used to model a problem for children. As the term manipulative suggests, children can handle and move pieces to solve the problem. For instance, Tag Bags® can be stacked, sorted in a variety of ways, and even tossed into hula hoops to learn the position words in and out. Other activities use Mix & Make Shapes™ to sort shapes by type or learn about area. Counters are also used frequently in Pre-K math. Moving objects to count helps children to understand that numbers represent quantities.

Guess and Check

We want children to know that trial and error is one way to approach a problem. We want children to understand that they should check each answer. Then if needed, they can repeat until they find the correct answer. Learning to identify a small quantity without counting is an example of this strategy. Children tell how many objects they think they have seen. They count the objects after answering to check. Eventually, their guesses become accurate.

Look for a Pattern

At this stage, children are identifying pattern units in the world around them. Activities in this teacher's guide have children using 4 Squares More Squares® to copy and grow simple patterns. Children sing and dance to "Pattern Dance" from the *Sing, Sound & Count With Me* CD for a truly multisensory experience. When children are good at recognizing patterns, they can reproduce them and use strategies to solve a problem.

Draw a Picture

Drawing pictures for problem solving is much like acting out problems or using manipulatives. It is used in the same way, to represent numbers or items in a problem. Pictures can be drawn when manipulatives are unavailable.

Talking About Math

Every day children naturally use math. Even though they are using math, many are unable to describe what they are doing. We want children to have many words: nouns for naming, verbs for moving and being, adjectives for describing, and prepositions for talking about position and time. Understanding words and what people say is essential for communication, learning, and thinking. The activities in this teacher's guide give children the words to describe their understanding. Then they can think and talk about the math concepts they are learning. As children explain their reasoning, a teacher can assist them in areas they may not understand or encourage them to explore further if they understand completely. Children also clarify their thoughts as they express their ideas and actions in words.

The *Numbers & Math Pre-K Teacher's Guide* helps children to talk through problems and engage in activities in a friendly, natural way. It highlights important math words in each activity and suggests questions to help children think through and verbalize their processes. The activities are engaging, so conversations flow easily. These conversations are a key part of math learning.

Now you may be thinking, I understand why children need to talk about math, but what about those concepts I'm not sure about myself? We've provided a glossary in the back of this teacher's guide to define math terms. The domain descriptions on page 9 and in each activity section introduction will help explain the skill areas. You don't need a lot of technical understanding to get children to talk about math. Remember these guiding questions to start conversations as your children participate in math activities.

- What are you making?
- How many do you need?
- What did you do here?
- What will you do next?

Talking about math involves listening, observing, asking questions, and helping children with the words they need. It depends more on your enthusiasm and encouragement and less on your math expertise. Enjoy the activities in this teacher's guide and talk with your children as you have fun together.

Getting Acquainted

You may be new to Get Set for School® and just using our Numbers & Math program. Or you may be a long-time Get Set For School user with all of our materials. Either way, you'll find our products and teaching tips to be fun and effective.

This teacher's guide is intuitive, easy to work with, and full of "aha" moments. Here's a quick orientation:

Domain Introductions – Each domain begins with a simple explanation of that learning area and lists the domain's specific skills in developmental order.

Multisensory Numbers & Math Products – Our hands-on products are at the heart of our success. Each product is introduced on its own page and in the domain where it is used most. We may use a product in other domains (sometimes even before it has been introduced). If you want to learn more about a specific product, here is a quick guide to where you can find those introductions:

Sing, Sound & Count With Me CD – page 26

Tag Bags® – page 28

I Know My Numbers – page 30

1-2-3 Touch & Flip® Cards – page 32

4 Squares More Squares® – page 70

Mix & Make Shapes™ – page 72

Discovery Play – Next to every product page is a page called Discovery Play. It describes how the product can be used for free play. Children need lots of time for free play. Discovery Play will develop their curiosity and understanding.

Other Get Set For School Products – You'll notice that we occasionally pull in one of our products from the Get Set for School Language & Literacy or Readiness & Writing programs because they work so well. If you don't have them, you may want to consider them, or use tools you already have in your classroom. For more information about other Get Set for School products, visit getsetforschool.com.

Activities – See pages 14–15 for a layout of our activity pages—the bulk of this teacher's guide. You will find the activities intuitive, child friendly, and effective for teaching important skills.

Resources – At the end of the book, we've collected useful resources that you will want to review during quieter moments when children are out of the classroom.

Teaching Guidelines – We are sticklers for developmentally based instruction (going from easier to harder), but there are many ways to mix in our products and stay true to your philosophy. We offer Teaching Guidelines to help organize your thinking/lessons.

Choosing Activities in This Teacher's Guide

The activities in this teacher's guide are organized by domain in developmentally progressive order. Here are four ways to choose activities within a domain:

1. **Choose what your children like.** Happy, engaged children are learning. It's fine to repeat favorite activities. Be sure to try the additional activities in the More to Learn section to vary or extend activities.

2. **Choose what your children need.** If you know children haven't been read to, or had experience with counting, blocks, shape sorters and puzzles, give them that exposure and those experiences at school. Be sure to use the ELL and Support sections to meet the needs of English language learners and children who need any extra help or support. The activities in the ELL and Support sections are motivating and hands-on so you can use them with children of different abilities and backgrounds.

3. **Choose what fits into your theme.** If you are using a theme approach, then turn to the Index on pages 192–193.

4. **Choose a range of activities.** Refer to the Teaching Guidelines on pages 178–185. These guidelines help you organize activities. However, you may order the activities to suit your children/teaching.

Even with the most thoughtful choice, some activities may surprise you when you try them. Here's what you may learn:

- **They're not ready.** There is no harm in trying. If an activity doesn't suit your children, you can just put it away for another day.

- **They're almost ready.** This could work with simple tweaking or a few more tries. Perhaps you'll choose a simpler related activity. Have them try a concept naturally through self-directed play. You may be pleasantly surprised as children discover things on their own.

- **They are ready.** This activity suits them perfectly and can be repeated. Be sure to use the exercises in the More to Learn section to vary and extend the activity.

Activity Design

Skill Timeline – This shows where each skill fits into the progression of skills. Large dot indicates skill being taught.

Title of Lesson - See next page for detailed description.

Skill – This is the skill being taught in this activity.

Description of the Skill – Important information about the skill is outlined for you.

Pictures – Illustrations or graphics give you a visual representation of the learning activity.

Look What We're Learning – The math benchmarks addressed by the main activity are listed. We've also included social-emotional and sensory motor benchmarks.

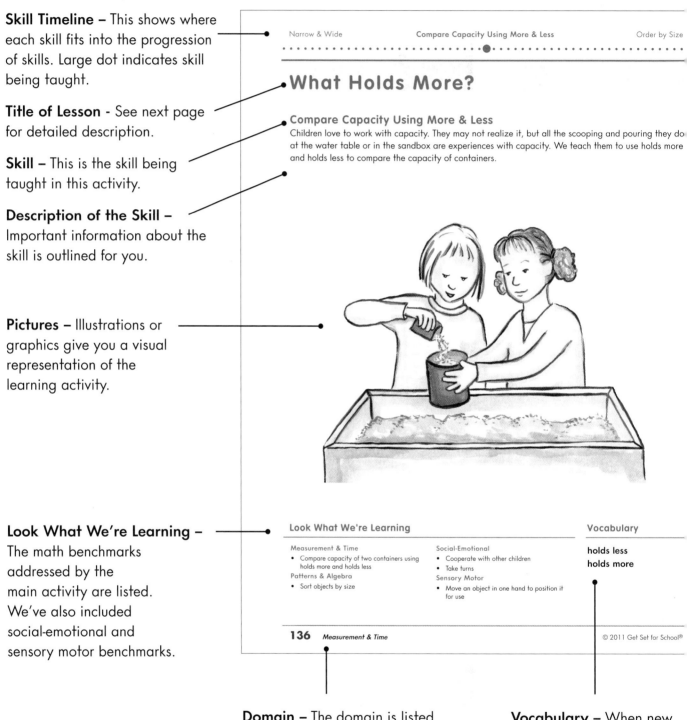

Narrow & Wide   Compare Capacity Using More & Less   Order by Size

What Holds More?

Compare Capacity Using More & Less

Children love to work with capacity. They may not realize it, but all the scooping and pouring they do at the water table or in the sandbox are experiences with capacity. We teach them to use holds more and holds less to compare the capacity of containers.

Look What We're Learning

Measurement & Time
- Compare capacity of two containers using holds more and holds less

Patterns & Algebra
- Sort objects by size

Social-Emotional
- Cooperate with other children
- Take turns

Sensory Motor
- Move an object in one hand to position it for use

Vocabulary

holds less
holds more

136   *Measurement & Time*   © 2011 Get Set for School®

Domain – The domain is listed at the bottom of the page to help you quickly identify what you are teaching.

Vocabulary – When new vocabulary is introduced, you will find this section. Definitions can be found in the glossary.

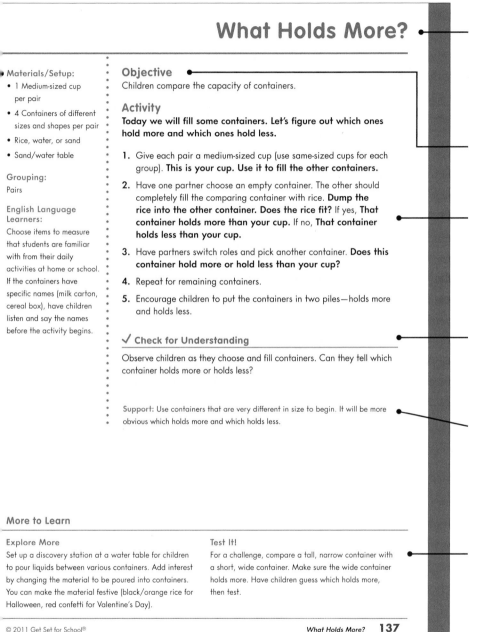

What Holds More?

Title of Lesson – Each lesson has a creative title describing the activity. You can also find the title of the lesson at the bottom of the page and before the skill description.

Materials/Setup:
- 1 Medium-sized cup per pair
- 4 Containers of different sizes and shapes per pair
- Rice, water, or sand
- Sand/water table

Grouping:
Pairs

English Language Learners:
Choose items to measure that students are familiar with from their daily activities at home or school. If the containers have specific names (milk carton, cereal box), have children listen and say the names before the activity begins.

Objective
Children compare the capacity of containers.

Objective – This is the purpose of the main activity.

Activity
Today we will fill some containers. Let's figure out which ones hold more and which ones hold less.

1. Give each pair a medium-sized cup (use same-sized cups for each group). **This is your cup. Use it to fill the other containers.**

2. Have one partner choose an empty container. The other should completely fill the comparing container with rice. **Dump the rice into the other container. Does the rice fit? If yes, That container holds more than your cup. If no, That container holds less than your cup.**

3. Have partners switch roles and pick another container. **Does this container hold more or hold less than your cup?**

4. Repeat for remaining containers.

5. Encourage children to put the containers in two piles—holds more and holds less.

Activity – The main activity tells you what to do in simple steps.

✓ Check for Understanding
Observe children as they choose and fill containers. Can they tell which container holds more or holds less?

Check for Understanding – This is an informal measure of a child's understanding of the main activity.

Support: Use containers that are very different in size to begin. It will be more obvious which holds more and which holds less.

Support – This section provides guidance on how to meet the needs of children who may have difficulty grasping the skill.

More to Learn

Explore More
Set up a discovery station at a water table for children to pour liquids between various containers. Add interest by changing the material to be poured into containers. You can make the material festive (black/orange rice for Halloween, red confetti for Valentine's Day).

Test It!
For a challenge, compare a tall, narrow container with a short, wide container. Make sure the wide container holds more. Have children guess which holds more, then test.

More to Learn – This section provides additional activities to vary and extend the main activity.

What Holds More? **137**

Materials/Setup – Suggested items you will need to gather for the main activity are listed. Materials for Support, English Language Learners, and More to Learn sections are listed within those sections.

Grouping – The recommended number of children for the main activity is indicated.

English Language Learners – Each lesson has a strategy to support English language learners.

Hands-on Products

We are excited to introduce our new math products. This is just a preview of each product. You can find an in-depth overview of each product within the various domains of this teacher's guide.

I Know My Numbers

I Know My Numbers brings numbers to life in your classroom. Interactive exercises will have your preschoolers building, rhyming, singing, and coloring their way to number recognition, number formation, number sense, and counting up to 10. Use the 10 booklets at school and send them home to engage families in home learning. Each new booklet engages children and their families in home learning. There are numbers to trace, pictures to color, nursery rhymes, finger plays, songs, and other number activities for your Pre-K children. See page 30.

4 Squares More Squares®

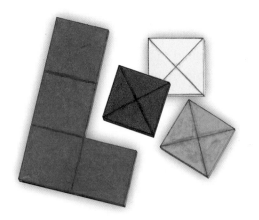

4 Squares More Squares makes geometry fun and easy. The colorful, chunky Big and Little Pieces appeal to children. They invite handling, arranging, and rearranging. They also stretch children's spatial awareness. Children use the pieces to make rectangles, squares, steps, and a variety of puzzles with the double-sided boards. The product grows with the child as the skills advance from counting and matching to sorting, patterns, graphing, and beyond! See page 70.

Tag Bags®

Tag Bags literally put math into children's hands. Multisensory activities teach preschoolers number recognition and number sense as they count, sort, measure, order, and build with Tag Bags. A variety of fasteners also promote important fine motor skills. Tag Bags activities are ideal for student- and teacher-directed learning. See page 28.

1-2-3 Touch & Flip® Cards

Children take on numbers with the flip of a card. **1-2-3 Touch & Flip Cards** feature tactile double-sided Animal Cards that entice children to trace and name numbers, count and sequence. "Flip" feature helps students check themselves and learn more easily. The smiley face ☺ in the top left corner helps children know that a number is right side up. Use for the whole class or center-based activities. See page 32.

Sing, Sound & Count With Me CD

Sing, Sound & Count With Me **CD** is filled with fun songs about math, literacy, and life both inside and outside the Pre-K classroom. In math, children learn about numbers, counting, shapes, and patterns all while singing to these upbeat, catchy songs. Your students will sing and dance their way to better math understanding and motor skills! See page 26.

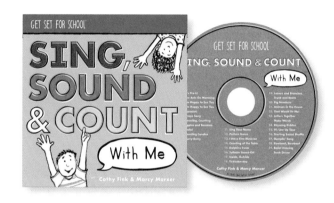

Mix & Make Shapes™

Geometry becomes relevant and accessible with **Mix & Make Shapes.** Multisensory activities help preschoolers practice problem solving and spatial reasoning with these brightly colored shapes. See page 72.

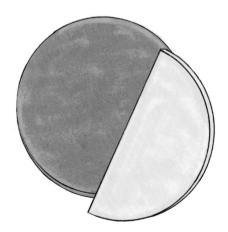

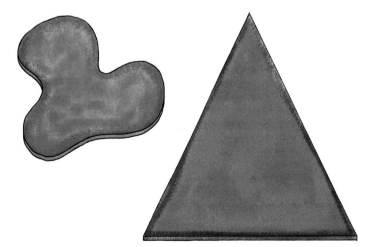

Pre-K Classroom and Children

Classroom Environment

Your classroom environment reflects many things about your teaching. An environment where children know where to locate items, what comes next in the day, and why they are participating in various activities provides structure and builds security. A cheerful, well-organized classroom helps you to teach effectively and allows your students easy access to the toys and materials that will help them grow and learn.

Social Environment

Provide a welcoming environment where all of the children feel accepted and part of the classroom community. Teaching and modeling ways to share, resolve conflict, and speak kindly to one another helps children learn and play with others. This in turn builds self-confidence.

Physical Setup

A good classroom setup promotes desired learning and behavior. Make sure the setup of your room offers a range of play and learning places. Set up the space and place materials intentionally to promote discovery play and learning. During teacher-directed times, make sure your students are situated so they can easily see and hear you. It's also important that children can easily participate in the activities during group exercises. Enhance learning in Pre-K with an organized space for teaching, working, and playing. Some ideas include:

- Dramatic play centers: kitchen, garage, store, etc., with related play materials
- Large spaces for independent, self-directed play and learning
- Spacious, active play area that allows children to move freely and participate in large group activities, such as dancing and whole group exercises
- Appropriately sized tables and chairs that allow you to position your students well for table activities
- Sensory and art centers, such as easels, sand, and/or water tables

Materials

There is a wide range of developmentally appropriate materials for Pre-K teachers to choose from. Every now and then, replace the materials in your room to offer new experiences. As you add things to your classroom, discuss the additions with your students. They can participate in how materials are set up and stored. Child-friendly labels with pictures/colors/words will help them to locate, sort, organize, and return objects to their proper places in the room. A good digital camera can come in handy during field trips or when you are designing classroom activities.

Walls

Walls can work for you because they can define spaces and support learning. They can also work against you if you put up every good thing because you'll have chaos and disorder. Use your walls actively and selectively.

Remember that the walls are for the children. Choose signs and labels to help children understand the purpose of print and begin to recognize letters and numbers. Instead of posting teaching standards or adult information on precious wall space, find an area where that information is easily accessible to adults.

Fine art posters, paintings, and arrangements add visual beauty to your room and increase cultural appreciation and awareness. You can vary permanent and changing displays to create interest. Display these at children's eye level to spark curiosity and wonder. Art shows emotion and introduces unfamiliar places and things. Talking about the art displayed around the classroom adds to children's background knowledge.

Audio

Every Pre-K classroom needs music. Music encourages movement and movement helps promote learning. Children benefit from music while they explore and play. Be considerate of children who are sensitive to noise. Use a low volume when you are asking children to focus on other work. Consider the wealth of our musical heritage. Include classical, jazz, country, and other genres—as well as children's music. Music is an important part of culture because the language, rhythms, and instruments introduce children to other times and places. Rhythmic and melodic patterns in music reinforce math understanding through multiple sensory channels.

Developmental Stages for Numbers & Math: Two- & Three-Year-Olds

Outside   Children learn math naturally at the playground. They discover diagonals on slides, height on ladders, and length on balance beams or curbs.

Inside   Blocks truly are the building blocks of math. Blocks teach counting and even geometry. There's math in books too. Say counting rhymes and sing number songs.

Daily Life   Taking a bath is body math when children notice fingers, toes, and one belly button. Getting dressed teaches math too—two socks, two shoes, two sleeves.

2-year-old

3-year-old

2-year-old

Counting and Numbers
- Counts aloud to 3; gives 1 object
- How old? Shows fingers, says "two"

Sorting and Comparing
- Knows familiar/new – food/clothes/toys, etc.
- Recognizes own things by attributes

Blocks and Building
- Builds tower of 6 cubes with precise release
- Builds 3-cube "train"

Toys and Puzzles
- With a few tries, places 4 shapes in a puzzle
- Plans direction for moving toys (forward/back, up/down)
- Stacks rings, nests 3+ cups by size

3-year-old

Counting and Numbers
- Counts aloud to 10; counts 3–5 objects
- How old? Shows fingers, says "three"
- Recognizes 1, 2, 3 in any order

Sorting and Comparing
- Sorts by size and attributes
- Compares big, bigger, tall, taller

Blocks and Building
- Builds tower of 10+ cubes

Toys and Puzzles
- Easily places 4+ shapes in a puzzle
- Strings 3 or more beads, connects train
- Fits toy parts together

Developmental Stages for Numbers & Math: Four- & Five-Year-Olds

Outside Young children learn math on the go. Signs have shapes and numbers to learn. Time and distance take on meaning. "Are we there yet?" Think of roads, parking lots, and construction sites as math opportunities.

Inside Children quickly outgrow stack and nest toys. They need more complex puzzles and games.

Daily Life Four- and five-year-olds like to help. Teach them to help set the table, sort the socks, and fold the towels to promote math skills. They learn about one-to-one correspondence and classifying by attributes, size, and shape.

4-year-old

Counting and Numbers
- Counts aloud to 20+; counts 6–10 objects
- Recognizes 1, 2, 3, 4, 5 in any order
- Counts backward from 10

Sorting and Comparing
- Knows clothing/shoes/possessions by person
- Sorts by attribute (color, shape) or use (to eat)

Blocks and Building
- Builds simple log/block house
- Builds very tall towers
- Copies 3-cube pyramid with cups or cubes

Toys and Puzzles
- Places and names many shapes in puzzle
- Puts ¼" pegs in simple design on board
- Follows an AB pattern for beads/cubes

5-year-old

Counting and Numbers
- Counts aloud to 50+; counts 10–20 objects
- Recognizes numbers up to 20 in any order
- Knows age-number meaning (5 older than 3)
- Counts on from a number (5, 6, 7, etc.)

Sorting and Comparing
- Sorts objects/animals with numbers
- Matches domino patterns

Blocks and Building
- Builds complex structures
- Carefully places pieces to balance tall tower
- Copies 6-cube pyramid with cups or cubes

Toys and Puzzles
- Assembles 12+ piece interlocking puzzles
- Uses color, shape, design to connect puzzles
- Puts shapes together for pictures

Meeting Individual Needs

Supporting English Language Learners

All young children need help to adjust to school, develop language skills, and learn. The English language learner needs help with all of that plus learn a new language and culture.

The beautiful thing about Pre-K is that all children, including English language learners, are acquiring the English language. Things are done differently in Pre-K for this very reason. The first step is to recognize that not all children live in homes where English is the primary language. In Pre-K, we can significantly impact children's learning with strategies to enhance a greater understanding of their world.

Throughout our *Numbers & Math Pre-K Teacher's Guide,* every lesson includes a strategy that you can use with your English language learners to support their learning. The strategies use hands-on experiences to promote vocabulary, understanding, and make connections between the home and school cultures. They also encourage children to speak, using their new words, and use multisensory experiences—music with movement, rhymes, and manipulatives.

Finally, when you work with English language learners, keep concepts or lessons in context. Allow children to touch and see the things you discuss. This is an important step to help them understand. As your children's language develops throughout the year, reflect on their drawings and dictated writings and see their language expand and grow.

Support Strategies for All Learners

Before children can access the curriculum, they need to be able to access the daily routines of the classroom and communicate their basic needs. Your classroom may have children who need a significant amount of support to get through the basics of the day for a variety of reasons, including limited English communication skills, limited home experiences, or developmental delays. Children also have varied interests, strengths, and backgrounds.

There are many strategies you will use and try throughout your teaching career. We like to focus on five strategies that are good for young children:

- **S**ocial
- **M**ultisensory
- **I**nstruction
- **L**anguage
- **E**valuation

Social: Make sure basic needs are being met and that emotional well-being is addressed. Children should be able to tell an adult when they need help, are hungry, or have to go to the bathroom. They need to feel safe and comfortable first, and then they are ready to learn. Give extra attention to be sure children can communicate their basic needs. All children can benefit from whole group learning when it is participatory. Use song, rhyme, or repetition to engage children. Children who don't have experience will naturally follow the others.

Multisensory: Children have different learning styles: auditory, tactile, visual, kinesthetic. Many learn best from a combination of styles. Include activities where no language is required. Use tactile, kinesthetic, and visual senses in directions as well as activities. When possible, use manipulatives to bring concepts to life. Our multisensory approach meets the needs of diverse learning styles and levels.

Instruction: You, as the teacher, are the gateway to learning. Teach with spirit to engage children and bring meaning to learning. Children can learn from the tone of your voice, your gestures, and interactions with others. Parents are their children's first teachers. Partnerships between parents and educators can greatly benefit student learning. Communication with parents is just as important as communication with your students. Whenever possible, use a child's experiences or background knowledge to make connections. Reach out to programs in your community that work with populations in your room. Network with school specialists.

Language: Use routines and repeated words or phrases for transitions. They have a comforting familiarity. Try to avoid long verbal explanations or questions by speaking in single words, short phrases, or short sentences. Speak slowly, even pausing between words. Use body language and facial expressions. Allow children to process the things you say. When you ask questions, give children ample time to process the information before you move on. Repeat lessons and activities. Young children can never get enough repetition.

Evaluation: You can use the Check for Understanding in each activity to evaluate the need for support. Each activity includes support for children who may not be ready for the current skill. It may include a prerequisite learning step or suggest breaking instruction into smaller steps. Expect children to understand, to have receptive language before they have expressive language. They are able to follow directions before they use words to ask and explain.

NUMBER & OPERATIONS
I Know How Many

Children show number sense long before they learn about quantities and number symbols (numerals). They know that they want more, even before they know the word "more." They learn that they have one head and two hands long before they can recognize the symbols 1 and 2. Families teach children to say their numbers and count their fingers, often through rhymes and songs. We help children understand the numbers that they say or sing by rote memory and give words and symbols for what they already know.

Solid understanding of how to count to find out "How many?" (quantity) is a foundation of all other math ideas. Children's math learning should help them develop number sense, discover relationships among numbers, and examine the properties of numbers (Burns, 2007). Children benefit from hands-on experiences with materials as they develop early concepts of numbers.

The activities in this domain allow children to build the following skills:

- **Demonstrate One-to-One Correspondence**
- **Count a Set of Objects**
- **Describe Cardinality**
- **Count in Any Order**
- **Make a Set**
- **Recognize Quantities Without Counting**
- **Use Ordinal Numbers**
- **Compare Sets**
- **Identify Numerals**
- **Connect Numerals to Quantities**
- **Write Numerals**
- **Label Sets**
- **Combine Sets**
- **Take Objects Away**
- **Share a Set Evenly**
- **Divide a Whole into Two Halves**

Below is the significant research for this domain. For additional Number & Operations resources, see the reference section at the end of this teacher's guide.

Howell, Sally, and Coral Kemp. 2009. "A Participatory Approach to the Identification of Measures of Number Sense in Children Prior to School Entry." *International Journal of Early Years Education* 17(1): 47-65.

Linder, Sandra M., Beth Powers-Costello. 2011. "Mathematics in Early Childhood: Research Based Rationale and Practical Strategies." *Early Childhood Education Journal* 39: 29-37.

Sadler, F. H. 2009. "Help! They Still Don't Understand Counting." *TEACHING Exceptional Children Plus* 6(1): Article 3.

Sing, Sound & Count With Me CD

The *Sing, Sound & Count With Me* CD includes 29 enjoyable songs, specifically chosen or written to reinforce key literacy and math concepts. Performed by Cathy Fink and Marcy Marxer, Grammy-winning songwriters and musicians, the CD features a variety of musical styles that capture your children's interest. Children happily sing them over and over and move to the captivating songs. The booklet included with the CD will help you with the words. Lyrics are also posted online on A Click Away.

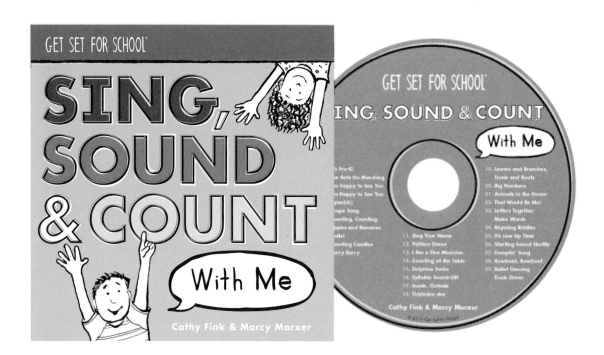

You'll like it because . . .

The songs link to activities in the *Numbers & Math Pre-K Teacher's Guide* and the *Language & Literacy Pre-K Teacher's Guide*. They introduce concepts or reinforce them. For example, "Shape Song" names shapes and also describes the characteristics of each one. Others help Pre-K children develop social-emotional skills, such as being able to transition smoothly from one activity to another, recognize emotions, demonstrate positive self esteem, and learn to cooperate.

Where you'll see it

Number & Operations	Patterns & Algebra
Geometry	Measurement & Time

Discovery Play with *Sing, Sound & Count With Me* CD

The best way to use *Sing, Sound & Count With Me* is to listen to the songs and read the lyrics on your own. Then play it in the background during free play to subtly introduce it to children. See which songs most appeal to all of you. Think about how the skills fit your plans. Here are some ideas. The math activities are in **bold.**

Track–Song	Suggested Activities
1 It's Pre-K!	Swing bent arms in time. Jump two times and wave hands for chorus.
2 The Ants Go Marching	**March around in groups. Hold up fingers to show numbers.**
3 I'm Happy to See You	Sit in a circle. Tap knees and clap with the beat.
4 I'm Happy to See You (Spanish)	
5 Shape Song	**Trace or hold shapes in the air. Show sides and corners.**
6 Counting, Counting	**Follow lyrics for movements while counting to 5.**
7 Apples and Bananas	Exaggerate mouth positions for sounds.
8 Smile	Make facial expressions for each verse.
9 Counting Candles	**Clap to the rhythm. Show fingers while counting.**
10 Hurry Burry	Make motions for each mishap.
11 Sing Your Name	Clap to the rhythm. Clap out the syllables in names.
12 Pattern Dance	**Dance and wave. Then follow lyrics for patterns.**
13 I Am a Fine Musician	Motion playing instruments. Clap syllables.
14 Counting at the Table	**Follow lyrics for motions, one person at a time.**
15 Dolphins Swim	**Swim and dive following lyrics for positions.**
16 Syllable Sound-Off	March around. Clap out syllables.
17 Inside, Outside	Hold index finger in front of mouth for quiet; Cup hands around mouth for loud.
18 Tickledee-dee	Sway to music. Point to child when name is sung.
19 Leaves and Branches, Trunk and Roots	Wave hands, sweep down arms and body, and pat floor.
20 Big Numbers	**Point to head for know. Shake head never. Act out last line.**
21 Animals in the House	Look around for animals. Motion animals' actions.
22 That Would Be Me!	Make motions for fly, grow, swim, and swing.
23 Letters Together Make Words	Grasp hands together. Say with cupped hands.
24 Rhyming Riddles	Walk in a circle for chorus. Stand still and nod to beat for riddles.
25 It's Line Up Time	Rhumba into line.
26 Starting Sound Shuffle	Point to self or children for call and response.
27 Dumplin' Song	**Act out question/answer with two groups. Count down with fingers.**
28 Rowboat, Rowboat	**Have children play the animals and climb into the boat.**
29 Ballet Dancing Truck Driver	Create motion for each occupation.

Tag Bags®

Tag Bags give Pre-K children a hands-on introduction to math. The set includes:

- 30 Tag Bags
  - Vibrant colors—red, yellow, blue, orange, green, purple
  - Fasteners to build fine motor skills—Velcro®, snap, hook, button, loop
  - Dots to promote one-to-one correspondence—one, two, three, four, five
  - Large, easy-to-read numerals—1, 2, 3, 4, 5
  - Pockets to hold Color Tags, name tags, counters, or other small objects
- 36 Color Tags
- Activity booklet—10 activities in the form of songs set to familiar tunes

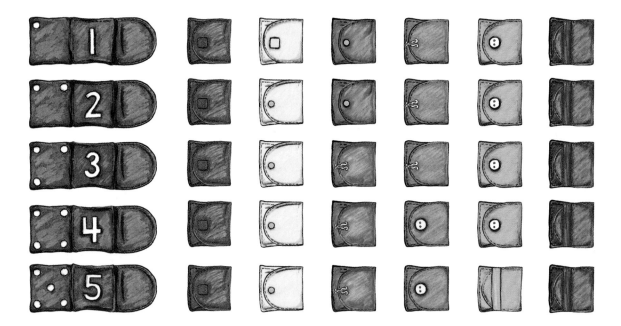

You'll like them because . . .

Tag Bags encourage children to open, close, count, sort, measure, and build. Tag Bags are versatile and easy to use, ideal for both student- and teacher-directed activities. Their applications can be simple or advanced. Primary and secondary color Tag Bags have different dot arrangements to show that the number of dots stays the same despite the arrangement. Color and fastener sorts reveal addition concepts. Use the activities in this teacher's guide, the activity booklet, or create your own great ideas to make the most of Tag Bags.

Where you'll see them

Number & Operations	**Measurement & Time**
Geometry	**Data Representation & Probability**
Patterns & Algebra	

Materials
- 2 Shoeboxes (cardboard or plastic)
- 30 Tag Bags
- 15 Color Tags

Box Suggestions
- 15 Tag Bags (3 color groups), 15 Color Tags (colors other than bags)
- 15 Tag Bags of one color, 15 Counters, Counter Cards 1–5

Discovery Play with Tag Bags®

Delightful teacher-led activities are in the activity booklet and this teacher's guide. Discovery Play with Tag Bags is different because it doesn't have set objectives or planned steps. The children are in charge of what they do and what they learn. The learning possibilities are endless, personal, and powerful. We want children to learn beyond what we teach them, to learn what they want to know. As children become familiar with the materials, follow teacher-directed activities, and watch others, their play will change. Ideally, they'll play with even more variety and creativity.

Including Discovery Play in Your Day

Children should have time to explore every day. It helps them generalize what they learned. Prepare Discovery Play boxes for children to borrow, use, put back together, and return. Children spread out and explore the contents of the boxes on a table or floor.

Procedure
- Set up the boxes. Let two to three children select each box.
- Step back—What can you do with these? Do not show children how to play.
- Sift through ideas—Resist taking over and teaching, but do make comments.
- Describe what you see with key vocabulary words: **How interesting! You made a row.** Ask guiding questions: **What did you do with your Tag Bags? Why did you put it here?** Let them tell you what they did. If they make errors, just say, **I see,** and then say it correctly.

Tag Bags are versatile so children can arrange, sort, and manipulate them. Children may do any of the following with the bags: toss; pile; line up; count; open; close; place open bags right-side up or upside down; sort by fastener or color; match counters to dots or bags to cards; and arrange randomly in a design, vertically, horizontally, by dots, in ascending or descending order, or in a pattern.

When they play, children may use any of these words: red, yellow, blue, green, purple, orange, square, open, close, corner, side, edge, fastener, pocket, row (horizontal), column (vertical), number, one, two, three, four, five, top, middle, bottom, left, right, right-side up, upside down, near, beside, between, on, order, same, different, more, fewer, same, or pattern.

I Know My Numbers

I Know My Numbers provides playful opportunities for children to explore number concepts.

- One booklet for each number 1–10
- Number writing practice
- Tips for parents and ideas for using household materials for manipulatives
- Number and counting activities for school or home

 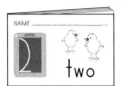   

    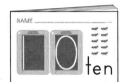

You'll like them because . . .

I Know My Numbers brings numbers to life in your classroom. Your preschoolers will actively build, rhyme, sing, and color their way to number recognition, number formation, number sense, and counting to **10**. The booklets reinforce number concepts in school and help parents understand and continue practice at home. The first booklet is all about the number one. Children learn how to write **1** and then they focus on "one" with rhymes, songs, coloring, finger plays, and math activities. Booklet 2 features the number two, and also reviews the number one. Each booklet builds on the one before, so children develop new skills over time as they get meaningful review. *I Know My Numbers* builds important math skills in a friendly way. Children will interact with the classic songs and rhymes, the fun activities, and the pictures to color.

Where you'll see them

Number & Operations   Patterns & Algebra

Geometry

Materials

- *I Know My Numbers* booklet 3
- 3 Caps
- 3 Cups

Discovery Play with *I Know My Numbers*

Young children need playful opportunities to explore number concepts, expand number knowledge, and practice writing numbers. *I Know My Numbers* introduces ideas for creative play around numbers and number concepts. The pages use real world and imaginary examples to encourage curiosity and show how numbers are part of everything.

Including Discovery Play in Your Day

The *I Know My Numbers* booklets are meant for use in both the classroom and in the home. There are many adult-directed activities and ideas for exploratory play. The inside front cover of each booklet lists ways to explore a specific number. There are creative places to look for numbers in your home and in the community. Sometimes nursery rhymes or classic stories are suggested that are built around the number. The following activity, which uses the *I Know My Numbers* booklet 3, is a great example for home or school.

Procedure - Caps in Cups

- Open booklet 3 to the page with the caps and cups (see example below).

- Provide these caps and cups to each child or to pairs. You can also arrange them as a center for children to explore. Children will naturally play with the materials. They may use the examples on the page to start, but will soon create their own ways to stack or arrange the materials.

- This activity can be done with any number of caps and cups.

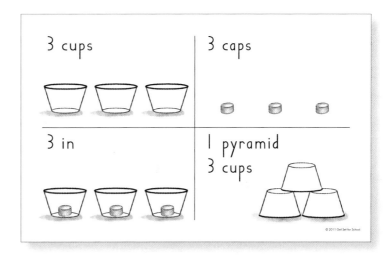

1-2-3 Touch & Flip® Cards

1-2-3 Touch & Flip Cards engage Pre-K children with hands-on number activities.
The set includes:

- 10 Animal Cards
  - Tactile numerals to touch and trace on one side
  - Animals to count on the other
- 10 Counter Cards
  - Squares to count on one side
  - Numerals to identify on the other
- Activity booklet with five activities

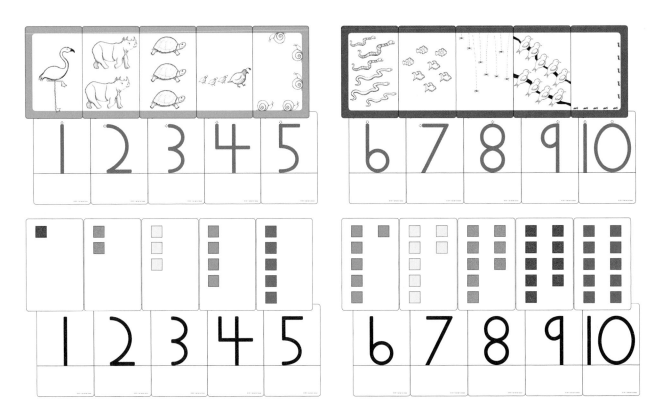

You'll like them because . . .

1-2-3 Touch & Flip Cards encourage children to name numbers, count with one-to-one correspondence, and form numbers. Tactile numerals on the Animal Cards promote tracing to introduce number formation. The familiar smiley face on the top left corner indicates when the card is right-side up. The Counter Cards feature squares in neat columns of five for organized counting. The shape and colors of the counters suggest a connection to Tag Bags® and 4 Squares More Squares®. You can use the cards in conjunction with these other products to reinforce counting with physical objects. Use the activities in this teacher's guide, the activity booklet, or create your own activities to make the most of the 1-2-3 Touch & Flip Cards.

Where you'll see them

Number & Operations

Materials

- 4 Shoeboxes (cardboard or plastic)
- 20 1-2-3 Touch & Flip Cards
- 55 4 Squares More Squares® Little Pieces
- 55 Counters

Packet Suggestions

- 10 1-2-3 Touch & Flip Animal Cards, 55 counters
- 10 1-2-3 Touch & Flip Counter Cards, 55 4 Squares More Squares Little Pieces

  OR

- 20 1-2-3 Touch & Flip Cards (Animal and Counter decks), 55 counters

Discovery Play with 1-2-3 Touch & Flip® Cards

1-2-3 Touch & Flip Cards encourage independent exploration of counting, matching, and sequencing. Discovery Play provides children opportunities to creatively play with the math concepts you've taught them while trying out their own ideas. The animal pictures, counters, and numerals on the 1-2-3 Touch & Flip Cards allow children to play at their own level, whether that's building a puzzle, counting with one-to-one correspondence, naming numbers, or even adding up big and small quails.

Including Discovery Play in Your Day

Children should have time to explore every day so they can generalize what they learn. Prepare Discovery Play boxes for children to borrow, use, put back together, and return. Children explore the contents of the boxes on a table or floor.

Procedure

- Set up the boxes. Let two to three children select each box.
- Step back—What can you do with these? Do not show children how to play.
- Sift through ideas—Resist taking over and teaching, but do make comments.
- Describe what you see with key vocabulary words: **How interesting! You arranged the counters to look like the ones on the cards.** Ask guiding questions: **What did you do with your Touch & Flip Cards? Why did you do it that way?** Let them tell you what they did. If they make errors, just say, **I see,** and then say it correctly.

You will notice that when children play with the 1-2-3 Touch & Flip Cards, they discover how to sequence numbers, sequence counters, sequence animals, count, trace numerals, sort by color, and match counters to squares or animals.

When they play, children may use any of these words: red, yellow, blue, green, purple, orange, square, row (horizontal), column (vertical), number, count, one, two, three, four, five, six, seven, eight, nine, ten, top, middle, bottom, left, right, right-side up, upside down, near, beside, between, on, order, same, different, more, fewer, same, pattern, flamingo, rhinoceros(es), turtle(s), quail(s), snail(s), snake(s), fish(es), spider(s), bird(s), ant(s), big, bigger, small, or smaller.

Table's Ready!

Demonstrate One-to-One Correspondence

Many children can count by saying numbers in order because they have memorized the words in the correct order. You can teach one-to-one correspondence by putting one cup with one plate. This concept prepares children to count with meaning and assign one number to each item.

Look What We're Learning

Number & Operations
- Match one-to-one up to 15 objects

Sensory Motor
- Look at hands and use visual cues to guide reaching for, grasping, and moving objects

Social-Emotional
- Cooperate with other children
- Take turns with peers
- Understand and follow classroom routines

Materials/Setup:
- Paper plates
- Paper cups

Grouping:
Pairs

English Language Learners:
Have a child pass out crackers for snack. Have him say, "One for you" with each cracker he gives.

Objective
Children learn one-to-one correspondence by setting a table.

Activity
Let's get the table ready for snack. We'll set one cup and one plate in front of each chair. Increase quantities when children are ready.

1. Set out plates and cups on a counter or side table. Assign a plate helper and a cup helper for each table.

2. **Plate helpers, here are the plates. Put one plate in front of each chair.**

3. **Cup helpers, here are the cups. Put one cup in front of each chair.**

4. **Plate helpers, is there one plate for each chair?** Helpers return extras or get more if needed.

5. **Cup helpers, is there one cup for each chair?** Helpers return extras or get more if needed.

You can repeat this activity daily.

✓ Check for Understanding

Have children repeat the activity, passing one napkin for each child. Are children showing one-to-one correspondence? Are they able to tell you how many napkins they passed out to each classmate?

Support: Use children's personal objects to help convey the concept—one jacket or hat for each child. Have three children at a time line up across from their jackets or hats. Each child puts on one item.

More to Learn

Socks and Shoes
You can vary the activity with socks and shoes. Set out three pairs of shoes. **How many socks go in each shoe?** Give a child socks to place one in each shoe. Increase the number as children understand the concept.

Counting Too
Take the activity one step further. Count the plates and cups as the children set them out. Emphasize that one number goes with each object and that the numbers do not match if there are fewer cups than plates.

Count With Me

Count a Set of Objects

Counting with one-to-one correspondence (or meaningful counting) is when children learn to say one number for each object in a set. Sometimes children lose their places, skip objects, or count objects twice. Children can count more accurately if they use a counting strategy, such as touching and moving objects as they count.

Look What We're Learning

Number & Operations
- Verbally count a set of 15 objects
- Match one-to-one up to 15 objects

Social-Emotional
- Cooperate with other children
- Take turns with peers
- Work with others to solve problems

Vocabulary

count

same

Materials/Setup:
- 2 Plates
- Counters

Grouping:
Pairs

English Language Learners:
Begin with one counter on the plate. Touch the counter and say **one**. Have the child repeat the number while moving the counter to the other plate. Repeat for each number. Children can count objects in their home language. They may already know the skill, but not the English words to communicate it. Read counting books and have children practice counting the objects in English.

Objective
Children count a set of objects.

Activity
Let's take turns counting.

1. Give each child a plate. Place four or five counters on one child's plate.

2. Have that child count each counter while moving it to her partner's plate.

3. Have the partner count and move the counters back to the first plate. **Did you count the same number of counters?**

4. If children count the same number correctly, add more counters. If children count different amounts, have them repeat the process.

✓ Check for Understanding

Observe children as they pass and count counters. Do they count correctly?

Support: Practice counting to 10 out loud. Take turns saying the numbers (teacher says **1**, child says **2**, teacher says **3**, child says **4**, etc.).

More to Learn

Count and Chant
Add interest with a chant. Play "Counting, Counting" (*Sing, Sound & Count With Me* CD, track 6). Demonstrate each verse. Repeat and have children join you in the chant and actions.

One in Each Box
Use pattern boards from 4 Squares More Squares® and counters. Children count as they place one counter in each square.

How Many in My Hat?

Describe Cardinality

"How many?" is a familiar question to children. They can find the answer by counting. The last number you say when you count a set tells how many, or the total. This concept is called cardinality.

Look What We're Learning

Number & Operations
- Recognize that the last number said is the total
- Verbally count a set of five objects
- Match one-to-one up to five objects

Literacy
- Listen to perform a task

Social-Emotional
- Cooperate with other children
- Participate in imaginary and dramatic play

Vocabulary

how many

in all

last

How Many in My Hat?

Materials/Setup:
- Hat
- Colored scarves, small objects, or counters
- Magic wand (optional)

Grouping:
Whole class

English Language Learners:
Play "Counting, Counting," track 6 on the *Sing, Sound & Count With Me* CD. Demonstrate each verse. Repeat and have children join you in the chant and actions. Have children raise one hand. **How many fingers?**

Objective
Children tell the total number of objects in a set.

Activity
How many scarves are in my hat? Count them as they appear.

1. Place three scarves in a hat so children can't see them.

2. Allow a few children to guess how many are in the hat. **Count the scarves as I pull them out of my hat.** Have a child help you wave the magic wand. Pull objects out slowly for effect, **1, 2, 3.**

3. **What was the last number we said?** (Three.) **Right. There were three scarves in my hat.**

4. Place the scarves in a line on the table. **Let's check. How many did we say? Let's count again—1, 2, 3.** Wave each scarf as the class counts. **The last number we said was three, so we have three scarves in all.**

✓ Check for Understanding

Invite children to be the magician. Hide several objects in the hat. Have them wave the wand and count the objects. Do they count the total number of objects correctly?

Support: Pass around the scarves or other objects to count. Have each child count them, one by one. **How many do you have in all?** (Three.) Pass them to the next child and repeat. Continue to practice with other quantities.

More to Learn

Tag Bag® Totals
Try using Tag Bags to vary the activity. Have children find blue Tag Bags in a mixed pile. Place the bags in a row as the children bring them to you. Count the bags together. **How many bags in all? The last number tells us—five.** Repeat with other colors.

Counting Buddies
Add interest by assigning children a counting buddy. Give each pair a category of classroom objects to count, e.g., trucks, chairs, crayons, etc. Have one partner count and the other repeat to check. Record their totals.

Count This Way & That

Count in Any Order

When children first learn to count, they may not understand the concept of totals. They may think that when objects are spread apart there are more than when the objects were close together. They may think that the order of counting affects the outcome. It takes repetition for them to understand that the arrangement or order of the objects does not change the total.

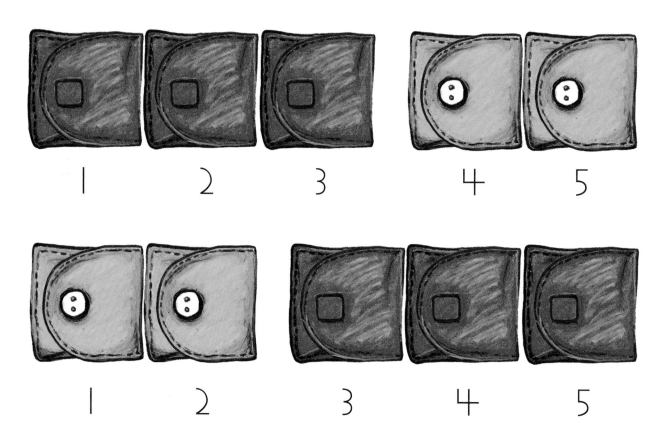

Look What We're Learning

Number & Operations
- Recognize that totals are not affected by order or arrangement
- Count a set of objects
- Recognize that the last number said is the total

Problem Solving
- Use manipulatives to find a solution

Social-Emotional
- Take turns
- Work together to solve a problem

Vocabulary

bottom
first
same
top

Materials/Setup:

- Tag Bags®
  - 2 Green
  - 3 Red

Grouping:

Pairs

English Language Learners:

Children can line up and count off. **How many did we count?** Start from the end of the line and count off. **How many did we count?** Now, have children switch places and count off again to show there is the same number of children.

Objective

Children recognize that order and arrangement of objects do not affect totals.

Activity

First, we'll count red and then green Tag Bags. Then we'll count green and then red. Will the number of Tag Bags be the same both times?

1. Toss three red and two green Tag Bags into the circle.

2. Have one child count the red Tag Bags first and then the green. Show her how to take each bag out of the circle as she counts. **How many are there?**

3. Invite another child to count the green Tag Bags first and then the red. Help him stay organized by moving each bag as he counts. **How many are there?**

4. **Did you get the same number? The number stays the same no matter what you count first.**

✓ Check for Understanding

Have children count the blocks in a tower. **Will the number be the same if we count from the top or the bottom?** Ask children how they know. Do children recognize that the order and arrangement of objects do not affect the total?

Support: Begin with two red Tag Bags and one green Tag Bag. Preview the activity by showing that two red Tag Bags and one green Tag Bag is three in all. Then show that one green Tag Bag and two red Tag Bags is three in all.

More to Learn

Line Count

Have children line up. Choose two children to be counting helpers. One helper counts the children in line from front to back. The other helper counts the children in line from back to front. Compare totals.

Tag Bag Dot Count

Show children a blue and a green number 3 Tag Bag. **The Tag Bags both say 3, but the dots look different. Do they really both have 3 dots?** Have one child count the blue Tag Bag dots and another child count the green Tag Bag dots. **Did they both have 3?**

Match My Number

Make a Set

Children make sets to show that they understand the meaning or value of numbers. When they consistently count the correct number of objects to match the number you say, you know they understand.

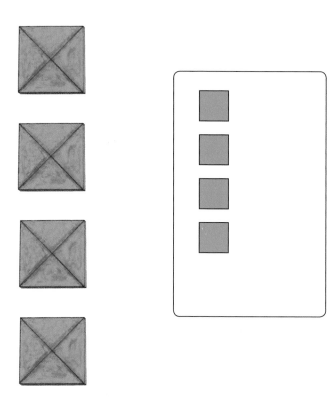

Look What We're Learning

Number & Operations
- Make or draw a set of objects to match a given number
- Match one-to-one up to 10 objects
- Verbally count a set of 10 objects

Sensory Motor
- Move an object in one hand to position it for use, placement, or release

Social-Emotional
- Cooperate with other children

Vocabulary

match

Match My Number

Materials/Setup:
- 1-2-3 Touch & Flip® Cards (Counter Cards)
- 4 Squares More Squares® Little Pieces or other counters

Grouping:
Small group (6 children)

English Language Learners:
Count and move all 10 counters with children so they become more familiar with the number words. Children can repeat the numbers as you say them. Then select a number for children to show with their counters.

Objective
Children make a set of objects to match the number given.

Activity
Use your counters to make the number I say.

1. Give 10 Little Pieces of one color to each child.

2. Demonstrate how to count and move each Little Piece up to 10 (2 columns of 5). Have children repeat with you.

3. Hold the Counter Cards number-side up in a fan. Invite a child to pick a card. Say the number to the group.

4. Have children count out the number of Little Pieces to match the card.

5. Flip the card to show the counters. Touch and count the squares as children touch and count their pieces.

✓ Check for Understanding
Observe children's sets and their counting. Can they match the set of objects to the number?

Support: Have children place counters directly on the Counter Card squares while counting. Remove the card and let them count again.

More to Learn

Set of Beads
Knot one end of laces used for stringing beads. Say a number 1–10. Have each of the children string that number of beads on the string. Check their counting by holding one end while they hold the other. Have them slide the beads as they count.

Flip, Check, and Pass
On the bottom of paper plates, write a number and the matching number of dots. Place the matching number of counters on each plate. Have children move their counters off the plate as they count. They can check their answers by flipping the plate.

Hide & Peek

Recognize Quantities Without Counting

The ability to recognize quantities up to three at a glance is an important skill for Pre-K children to develop. When children can identify a number without counting it, they understand and have internalized it. With this skill they move easily to performing operations with numbers. Children need many opportunities to see and identify the quantity in a set.

Look What We're Learning

Number & Operations
- Tell at a glance how many are in a set up to three without counting
- Make or draw a set of objects to match a given number
- Verbally count a set of five objects

Social-Emotional
- Imitate teacher's movements
- Take turns with peers

Vocabulary

how many

Materials/Setup:

- For each pair:
  - Shoe box
  - 5 Identical small objects

Grouping:
Pairs

English Language Learners:

Review the question
How many are there?
Pantomime how to count and answer with the number. Repeat the activity several times, and then pair the children with others in the class.

Objective
Children recognize how many without counting.

Activity
Tell me quickly how many you see.

1. One child in each pair is the hider and the other child is the peeker.

2. Demonstrate the hider's job. **I have three blocks.** Show children the blocks in a row and count them together. Cover them with the box, tipping it so you can reach beneath. For the first time, do not change the number of blocks.

3. Lift the box. **How many are there?** After a few seconds, cover the blocks. Invite children to answer. Repeat until children have the idea.

4. Pass out boxes to the hiders. Have them place their objects, lift the box, and ask, **How many are there?**

5. Give a signal for hiders to cover the blocks. Allow all peekers to answer. Hiders tell peekers if they are correct.

6. After a few turns, switch roles. Increase number of objects if children can recognize quantities easily.

✓ Check for Understanding

Observe as children tell how many cubes they see. Are they able to recognize how many cubes without counting?

Support: Count objects together. Have children make sets of numbers 1–5 in random order. Try the activity without the box several times. Add the box when children show understanding.

More to Learn

Tag Bag Flash
Show an open Tag Bag 1–3 with the flap covering the number. Ask children to quickly tell the number of dots.

Match My Dots
You can help children practice this skill with dominoes. Have children sit in pairs facing each other. Give each child several dominoes. One child shows one end of a domino and tells how many dots. The other child matches it and repeats the number. They trade roles and repeat.

Tag Bag Line Up

Use Ordinal Numbers

Up to this point, children have been working with counting numbers. Ordinal numbers are different. First, second, and third sound different and are used differently—to show position in a line. Children practice using ordinal numbers by putting things or themselves in order.

first

second

third

Look What We're Learning

Number & Operations
- Say an object's position in a line using ordinal numbers

Social-Emotional
- Manage transitions well
- Cooperate with other children

Sensory Motor
- Use fingers to open and close fasteners

Vocabulary

first

second

third

Materials/Setup:
• Tag Bags®

Grouping:
Small group (9), whole class when repeated

English Language Learners:
Have a child open the red Tag Bag 1. Say, **one.** Have him repeat, "one." **You are first.** Have him say, "I am first." Repeat with other numbers.

Objective

Children tell a person's position in a line using ordinal numbers.

Activity

Let's find out who will be first, second, and third in our lines.

1. Place Tag Bags 1, 2, 3 in three colors in a basket. Invite children to take a Tag Bag from the basket. **Find the other children with Tag Bags that match your color. Now open your Tag Bag.**

2. **If your Tag Bag has one dot, you are first in your color line. Stand here.**

3. **If your Tag Bag has two dots, you are second in your color line. Stand behind the first person.**

4. **If your Tag Bag has three dots, you are third in your color line. Stand behind the second person.**

5. Have first children take a big step forward saying, "I am first." Have second children take a step forward saying, "I am second." Have third children take a step forward saying, "I am third." Repeat as desired.

✓ Check for Understanding

Observe children as they say their ordinal positions. Can they say their position using ordinal numbers?

Support: Count three children sitting—**1, 2, 3.** Then have them stand in a line. Say the ordinal numbers for their positions—**first, second, third.** Have each child tell her position.

More to Learn

Looking for Lines

Another way to teach ordinal numbers is to have children look for lines of objects or people around the classroom or school. Have them identify the positions in line by saying the ordinal numbers.

Line Up Time

Play "It's Line Up Time," track 25 from *Sing, Sound & Count With Me* CD and invite children to line up. When they are in line, walk by them, saying the ordinal number for each place in line. Then have each child say his position in line.

More or Fewer?

Compare Sets

Children always want to have more, but do they really know how to make a comparison? Children learn more and fewer by counting and comparing sets or groups of things. It helps to line up objects so they see longer or shorter lines and understand which set has more or fewer.

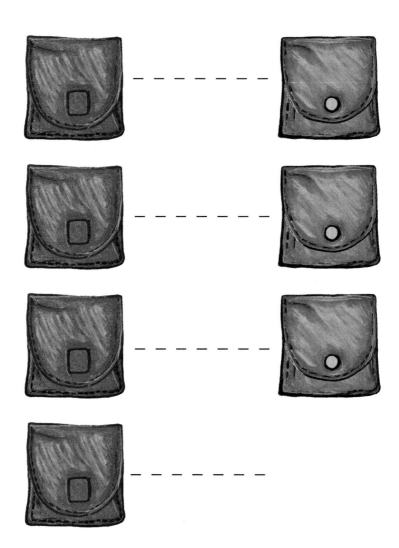

Look What We're Learning

Number & Operations
- Compare sets of objects using more and fewer
- Verbally count a set of five objects
- Recognize that the last number said is the total

Social-Emotional
- Cooperate with other children
- Work with others to solve problems

Vocabulary

fewer

more

More or Fewer?

Materials/Setup:
- Tag Bags®
  - 5 Red
  - 5 Blue

Grouping:
Pairs

English Language Learners:
Repeat the activity using varying numbers of different color Tag Bags. Have children say, "I have **more** red Tag Bags" and "I have **fewer** blue Tag Bags."

Objective
Children compare sets to see more and fewer.

Activity
Do you have more blue Tag Bags or red Tag Bags? Let's compare to find out.

1. Children work in pairs. One has red Tag Bags. One has blue Tag Bags.

2. Children take turns placing Tag Bags side by side, one red Tag Bag with one blue Tag Bag.

3. **Which set has pieces without partners? That set has more. The set with no extras has fewer.**

4. **Is that really true? We can count to be sure.** Invite children to count each line with you. **The set with the bigger number has more. The set with the smaller number has fewer.**

5. **Repeat, changing the numbers of red and blue Tag Bags.**

✓ Check for Understanding

Observe children as they compare red and blue Tag Bags. Do children know which has more? Fewer? Are they using the vocabulary **more** and **fewer**?

Support: Give two children two to six beads. Have them place their beads, one opposite the other, in separate rows of an egg carton. Help them identify the longer row. Repeat the words **more** and **fewer** to describe the amount of beads.

More to Learn

Count Each Set
Another way to compare is to count each set. Give each child a 1-2-3 Touch & Flip Animal Card. Form pairs. Have children count the animals on their cards and compare numbers with their partners. **The higher number has more animals. The lower number has fewer animals.**

Let's Add More
Add a challenge by increasing the number of objects in each set to 5–10 objects or 11–15 objects.

Name That Number

Identify Numerals

Children begin to count without understanding the meaning behind numbers. After they learn that numbers represent quantities, they learn to associate quantities with written symbols (numerals). They learn how the symbols look and how they are positioned on a page.

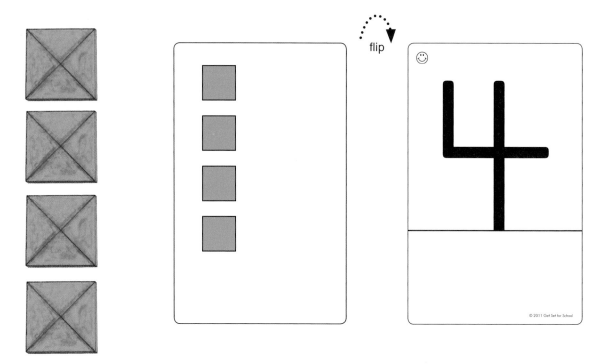

flip

Look What We're Learning

Number & Operations
- Identify written numerals and position correctly
- Verbally count a set of 10 objects
- Recognize the last number said is the total
- Make or draw a set of objects to match a given number

Social-Emotional
- Cooperate with other children
- Take turns with peers

Vocabulary

flip

Name That Number

Materials/Setup:
- 1-2-3 Touch & Flip® Cards
  Counter Cards 1–5
- 4 Squares More Squares®

Grouping:
Small group

English Language Learners:
Show a Counter Card (counter side up). Point and count the squares together. Have the child flip the card and repeat the number. Continue with the other numbers. Children can also say the numbers in their home languages to make connections.

Objective
Children position numerals right-side up, recognize, and name written numerals.

Activity
How do our numbers look? Our Touch & Flip Cards will help us find out.

1. Have child choose a Counter Card. **Count the squares. Your card has 4 squares.**

2. Have another child count out the matching number of little pieces. **You matched. You have 4 counters.**

3. **Let's see what 4 looks like.** Have third child flip the Counter Card. **Yes that's it, it's 4!**

4. Continue this pattern with the remaining cards.

✓ Check for Understanding

Have children choose a quantity of counters. Match counters to their numeral cards. Are children matching correctly? Are they positioning their numerals right-side up?

Support: Place Tag Bags® of one color in a bowl. Invite children to choose a Tag Bag. Have children open their Tag Bags one at a time. Have them count the dots and say the number, showing it to the group.

More to Learn

Counters in a Cup
Set out Counter Cards with a counter on top of each square. Place a cup next to each card. Have children count the counters into the cup. Then flip the card to see the numeral. Guide children to position the card correctly and air trace the number.

Continue On
You can extend this activity by continuing with numbers 6–10.

Show Me the Number!

Connect Numerals to Quantities

After children learn that numbers represent quantities, they learn to associate those quantities with written symbols (numerals). As the connections in their minds become stronger, children think of the quantity as soon as they see the numeral or number symbol.

Look What We're Learning

Number & Operations
- Connect numerals to quantities they represent
- Verbally count a set of 10 objects
- Recognize that the last number said is the total
- Make or draw a set of objects to match a given number

Social-Emotional
- Cooperate with other children

Sensory Motor
- Move fingers to show age/number and for finger plays

Show Me the Number!

Materials/Setup:
- 1-2-3 Touch & Flip® Cards
  Counter Cards 1–5

Grouping:
Small group

English Language Learners:
Use real objects or counters to count the numbers. If possible, use manipulatives because, when children can touch and move objects, the concepts become real.

Objective
Children connect numerals to the quantities they represent.

Activity
What does my number mean? Can you show me?

1. **I'm going to hold up a number. You hold up fingers to show what my number means.**

2. Give all the children time to hold up the correct number of fingers.

3. **My number is 3.** Everyone repeats the number and counts fingers together.

4. Continue through all the numbers.

✓ Check for Understanding

Show a Counter Card (number-side up). Have a child set out counters to show the number. Invite him to flip the card. Check by placing the counters on the squares and counting out loud. Do children connect numerals with matching quantities?

Support: Hold up one finger or two fingers to demonstrate. Help children place their thumb to show one or two fingers as needed. Name body parts and have children hold up fingers to show how many they have: **Head** (1) **Eyes** (2). Continue with other body parts. Say tail, horns, or wings just for fun.

More to Learn

Clap a Number
Add interest by clapping for the number shown. Show the number and let the children try to clap that number of times. Then say the number. Clap and count together. Try substituting another action, e.g., jumping, patting your head, and so forth.

Play a Game
Board games are another great way to practice this skill. Use a spinner or number cards to tell how many spaces to move. Children count the spaces to move their pieces.

Wet-Dry-Try

Write Numerals

Writing must be taught explicitly. Children form letters and numbers correctly without reversals.
Use the Get Set for School® strategies to teach capitals and numbers.

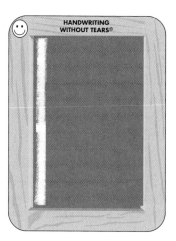

Teacher's Part
Demonstrate correct
number formation

Child's Part

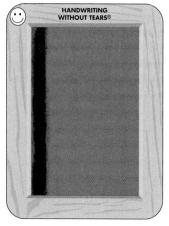

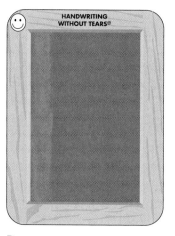

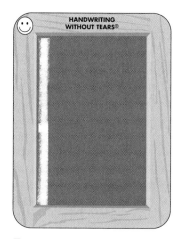

Wet
Trace 1 with wet sponge,
then finger

Dry
Dry 1 with crumbled
paper towel

Try
Write 1 with chalk

Look What We're Learning

Number & Operations
• Write numerals up to 10

Sensory Motor
• Develop correct pinch grasp
• Use index finger to trace numbers on slate

Materials/Setup:
- Slate
- Little Chalk Bit
- Little Sponge Cube
- Small square of paper towel

Grouping:
Small groups

English Language Learners:
Use Wood Pieces to familiarize children with the formation language— Big Line, Little Line, Big Curve, Little Curve. Demonstrate the number with the Wood Pieces on the blue Mat. Have children practice the language by forming their first initials. For more information about using Wood Pieces, see our *Readiness & Writing Pre-K Teacher's Guide.*

Objective
Learn correct formation for numerals up to 10.

Activity
You know what the numbers look like. Now we're going to write them.

1. **We'll start with 1. I will write the number 1. One starts in the starting corner with a smiley face. Big Line down.** Draw **1** with the chalk along the left edge of the slate.

WET
2. Have the child wet a sponge cube and squeeze it out. **Now trace the 1 with your sponge. You can wet your finger and trace it again.**

DRY
3. **Now crumple your paper towel. Dry the 1 with your towel a few times.**

TRY
4. **Now you can use the chalk to write a 1 in the same place. Start in the starting corner. Big Line down.**

5. Practice writing **1** in small groups with prepared slates.

6. After children can write **1** easily, continue with **2**. Refer to the Number Formation Chart at ***getsetforschool/educators/classroomextras*** for formation examples and language.

✓ Check for Understanding

Watch children as they trace the number. Does the number start at the top? Is the number formed with the right sequence of strokes?

Support: Use the 1-2-3 Touch & Flip® Animal Cards (number side up). Say the formation words as children trace over the tactile number with their index finger.

More to Learn

I Know My Numbers
Practice number formation with *I Know My Numbers.* First, children trace **1** on the cover with their index finger. **Big Line down.** On the next page, children trace **1**, using a Flip Crayon. Check their grip. Guide them to stabilize the booklet with their helping hand.

Door Tracing
Place a large smiley face in the top left corner of the classroom door. When children are lined up at the door, demonstrate a number, saying the formation language and using the door as an imaginary slate. Have children trace the number before they pass through the door.

9, 10, A Big Fat Hen

Review Writing Numerals

Teach children to write numbers in numeric order. Numbers 1 through 7 all start in the top left starting corner so the formation habits are easy to learn. Introduce every new number with multisensory activities. Progress from large multisensory teaching to tracing and eventually to independent number writing. Eight, nine, and ten are taught last.

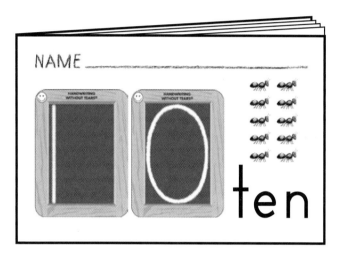

Look What We're Learning

Number & Operations
- Write numerals up to 10
- Connect numerals to quantities they represent
- Verbally count a set of 10 objects

Sensory Motor
- Use fingers to hold crayon and other objects
- Use helping hand to stabilize papers

Materials/Setup:

- *I Know My Numbers,* booklet 10

- Flip Crayons®

Grouping:

Small group, whole class

English Language Learners:

Review the names of the Wood Pieces used to form numbers: Big Line, Little Line, Big Curve, and Little Curve. Preview the instructional words we use to teach crayon placement and number formation: smiley face, arrow, and aim. Use consistent language and live demonstrations to build familiarity.

Objective

Review correct formation for numerals up to 10.

Activity

Let's review our numbers with a nursery rhyme. This is the last booklet in the set. Use the booklets in order, and follow the suggestions in each.

1. Read the nursery rhyme aloud. Invite children to turn pages in their own booklets to follow along.

2. Teach the first page. Show children how to hold the crayon correctly. **Let's review 1 and 2. For 1, aim your crayon at the smiley face and make a Big Line down.**

3. **For 2, aim your crayon at the smiley face and make a Big Curve and a Little Line.**

4. **You wrote 1 and 2, now you can color the shoe.**

5. Practice writing in the booklet over a two-week period. When finished, send it home to families.

✓ Check for Understanding

Watch children as they trace the numbers. Do they start in the correct spot? Do they write the numbers using the right sequence of strokes? Do they use a correct crayon grip?

Support: Use Wet-Dry-Try with the Slate Chalkboard to review number formation and to teach proper grip.

More to Learn

Fun Practice

Practice number formation with the *My First School Book* activity book.

Without a Trace

Use 3" x 8" strips of paper. Put a smiley face in the top left. With each child, demonstrate how to write the number on the strip. Have each child write numbers independently.

Write How Many

Label Sets

Children have learned to write their numbers and know what numbers mean. When they count and label sets, they practice these skills together.

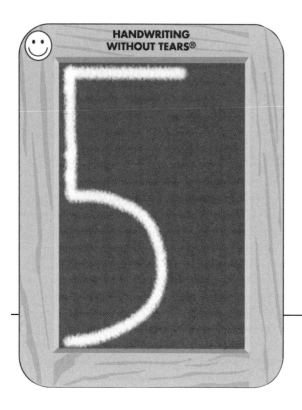

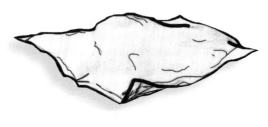

Look What We're Learning

Number & Operations
- Count and label sets up to 10
- Write numerals up to 10
- Connect numerals to quantities they represent

Problem Solving
- Draw a picture to help solve a problem

Social-Emotional
- Manage and handle transitions well and without incident

Materials/Setup:

- 5 Slates
- 5 Little Chalk Bits
- 5 Little Sponge Cubes, slightly wet
- 5 Small paper towel squares
- 5 Paper cups
- 15 Plastic bottle caps

Grouping:

Small group (five children)

English Language Learners:

Have children say the numbers out loud as they count. Assist with pronunciation as needed. As you give directions, pick up each of the materials and name it. Children can repeat the names aloud. Act out the process as you are explaining to show children what to do.

Objective

Children count and label sets.

Activity

Let's write numbers to tell how many caps.

1. Set up five stations on tables around the room. Each station should have: a slate with a number 1–5, a cup filled with the matching number of caps, a Little Chalk Bit, a Little Sponge Cube, and a small paper towel square.

2. Have children count the caps to make sure the cup has the correct number. On the back of the Slates, show children how to make dots to show the number on the Slate.

3. Have children repeat the process at another station.

✓ Check for Understanding

Observe children as they do Wet-Dry-Try, count, and trace the caps. Can they count and label sets?

Support: Spread the 1-2-3 Touch & Flip® Animal Cards (number-side up) on the table. Have children count the number of caps in each cup and choose the correct card to label it. Guide them in tracing each tactile number.

More to Learn

More Labels

Extend the activity by including numbers 6–9.

Clap & Trace

Have a child sit in a chair facing away from the class. He claps a number 1–10. The class repeats and counts each clap. Then the class traces that number in the air as you demonstrate.

Count Them All

Combine Sets

When children can count objects in a set, they put more objects into their sets and count again. They don't realize that this process is addition. We clarify the steps, but the process is natural.

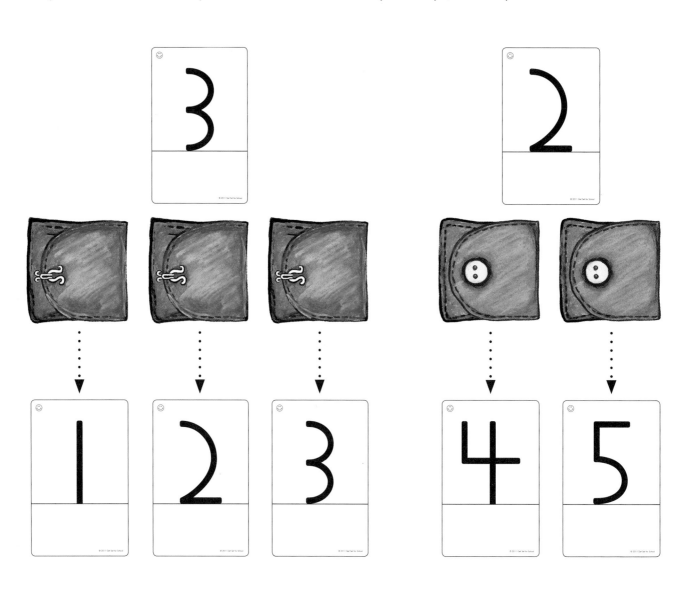

Look What We're Learning

Number & Operations
- Combine sets to learn how many in all by counting
- Connect numerals to quantities they represent

Problem Solving
- Use manipulatives to find a solution

Social-Emotional
- Cooperate with other children
- Take turns with peers
- Work with others to solve problems

Vocabulary

counting on
how many
in all

Materials/Setup:
- Blue and orange Tag Bags®
- 1-2-3 Touch & Flip® Cards 1–5

Grouping:
Small group

English Language Learners:
When children count Tag Bags in the activity, make a circle motion around each color group of Tag Bags as you repeat the numbers. **Three blue. Two orange.** Motion again around the entire line of Tag Bags. **All.** Count all of the Tag Bags. Pause after counting the first set then count on. **Three and two makes five in all.** Have children repeat as you point to each group of Tag Bags.

Objective
Children combine sets of objects to find out the total by counting.

Activity
How many do we have in all? Let's count to find out.

1. Invite two children to sit with you, one on each side. Others will watch and count with you.

2. Give the child on your right some blue Tag Bags. Count them aloud as a class. Have him repeat the total. Give him the corresponding number card.

3. Give the child on your left some orange Tag Bags. Count them aloud as a class. Have her repeat the total. Give her the corresponding number card.

4. **How many in all?** The first child places Tag Bags in a line, counting aloud. The second child adds Tag Bags to the row, counting on and continuing to count from where she first left off.

5. Review the numbers for each color as you show the number cards.

6. **Three** and **two** makes **five** in all. Have children repeat the sentence with you.

✓ Check for Understanding

Repeat with other Tag Bag colors. Are children able to count how many Tag Bags in all?

Support: Start with one Tag Bag of each color. Put the Tag Bags together. **How many in all?** Say together, **There are two in all.** Continue with two and one, two and two, and so on, increasing by one each time.

More to Learn

Place & Count
You can vary this by using 4 Squares More Squares® Little Pieces and the 2" x 3" square Pattern Board. Have two children place pieces of two different colors, first counting their own pieces and then the total.

"Rowboat, Rowboat"
Listen to "Rowboat, Rowboat," track 28 on *Sing, Sound & Count With Me* CD. Say the lyrics very slowly and act out each part.

How Many Are Left?

Take Objects Away

Children love to do this activity when they are eating small snacks. They know how many snacks you gave them to start with and want to know how many they have left after eating a few. As with addition, we are confirming the steps to something children do naturally.

Look What We're Learning

Number & Operations
- Take objects away from a set to find out how many are left by counting
- Make or draw a set of objects to match a given number
- Recognize that the last number said is the total
- Verbally count a set of 10 objects

Problem Solving
- Use manipulatives to find a solution

Vocabulary

take away

left

How Many Are Left?

Materials/Setup:
- Fish crackers, grapes, or other small snacks

Grouping:
Small group

English Language Learners:
Preview the activity with children. Repeat words and have them practice saying, **One for you, one for me,** so they feel more comfortable in the activity with their partner. Teach with spirit to engage your students and bring meaning to learning. Children can learn from the tone of your voice, your gestures, and interactions with others.

Objective

Children take objects away from a set and count to find out how many are left.

Activity

How many are left? We can count to find out.

1. Pass out six crackers to each child. **I have six crackers.** Have children repeat.

2. Have everyone pick up one cracker. Say together, **I take away one cracker.** Invite children to eat one cracker.

3. **How many are left?** Have children count their remaining crackers and tell how many are left. Say together, **I have five crackers left.**

4. Repeat, first eating two crackers to leave three, and then eating three crackers to leave zero.

✓ Check for Understanding

Have a child count all the counters in set of 8. Then count out three counters and cover them with your hand. **How many are left?** Do this several times, changing the number of counters covered. Are children able to say how many are left?

Support: Start with five crackers. Take away one at a time. Count how many are left each time.

More to Learn

"Dumplin' Song"
Listen to "Dumplin' Song," track 27 on *Sing, Sound & Count With Me* CD. **What happens every time the singer asks the questions?** (One more dumplin' is gone.) Act out the song, inviting a child to take a Tag Bag® dumplin' each time.

Take Away More
To add a challenge, increase the number of counters to 10 and take away 5–9 counters.

Everyone Gets the Same

Share a Set Evenly

We teach our children to share. Sharing teaches early division concepts. Sometimes we divide a group of things. To teach children division, we have children pass out objects one by one until the objects are gone.

Look What We're Learning

Number & Operations
- Share a set of objects evenly with two or three classmates
- Recognize that the last number said is the total
- Verbally count a set of 10 objects

Problem Solving
- Use manipulatives to find a solution

Social-Emotional
- Cooperate with other children
- Work with others to solve problems

Vocabulary

same

share

Everyone Gets the Same

Materials/Setup:
- 1 Tag Bag®
- 4 Buttons (or other small objects)

Grouping:
Pairs

English Language Learners:
Emphasize the starting number. Say **one** each time you pass an object. Have children count their objects. Say the total for each of the children and tell them that **these are the same.**

Objective
Children share a set of objects evenly with two or three classmates.

Activity
Let's see how we can share a group of things.

1. **There is something in this Tag Bag for us to share. Open it. What did you find?**

2. **Share the buttons with your friend. Here's a good way to share: one for you, one for me, until they are gone.**

3. **How many did you get?** Have both children count. **Do you have the same? That's sharing.**

4. **Repeat, using even numbers of buttons—6 or greater.**

✓ Check for Understanding

Invite children to help pass out snacks or pieces of paper to a small group of friends. Ask them how to make sure children get the same number. Are they able to group objects evenly?

Support: Begin with two buttons. Emphasize that each person gets the same number. Move on to four buttons.

More to Learn

Share Other Things
Use other objects such as pennies, counters, or beans to vary this activity.

Share with More Friends
For a real challenge, increase the group size to 3 children and 6 or 9 buttons or four children with 8 or 12 buttons.

Half & Half

Divide a Whole into Two Halves

Another way to divide is to cut or break one whole object into two equal pieces. Children are used to seeing this when we cut fruit or a large cookie to share.

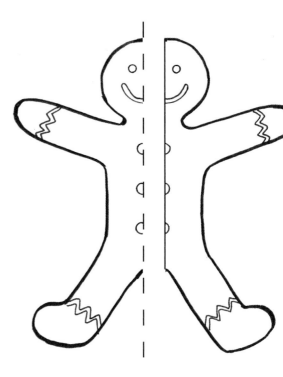

Look What We're Learning

Number & Operations
- Divide one whole object into two equal pieces

Problem Solving
- Use manipulatives to find a solution

Social-Emotional
- Cooperate with other children
- Work with others to solve problems

Vocabulary

half
middle
whole

Materials/Setup:

- *The Gingerbread Man*
- Flip Crayons®
- *I Know My Numbers* booklet 2
- Gingerbread man cookies or Gingerbread Cutout from A Click Away

Grouping:
Pairs

English Language Learners:

Show a whole orange (peeled). Emphasize **whole** and motion with your hand over the whole orange. Break the orange in **half.** Point to each half and say **half.** Hold them together and say **whole.**

Objective

Children divide one whole object into two equal pieces.

Activity

We can share by cutting one whole thing into pieces.

1. Read *The Gingerbread Man.*

2. Give each child a gingerbread man cookie or a Gingerbread Cutout. If you are using the cutout, have children color.

3. **Pretend you and a friend caught the gingerbread man. You want to share him. How can we cut him so you can share him equally?** Use a knife or scissors to show children how to cut the gingerbread man in half. Cut from head to toe so that pieces are symmetrical.

4. Give each child an *I Know My Numbers* booklet 2 to take home. Encourage them to share the story with their families and show how to cut the gingerbread man in half.

✓ Check for Understanding

Observe children as they cut their cookies/cutouts in half. Do they cut from head to toe? It will be a challenge for many to cut evenly, so celebrate effort.

Support: Show items already divided into two pieces. Some have equal pieces and others have one piece bigger than the other. Show each example and ask, **Is this good sharing?** Pretend to give them to two children. **Good sharing has two pieces the same size.**

More to Learn

Cookie Cutting

Have children form balls with play dough and flatten them into circles. Have them cut the circles in half to share.

A New Fold

Give each child a paper square to add complexity. Ask children if they can share it by making triangles. Show them how to fold the paper diagonally. Help children cut their squares into two triangles.

GEOMETRY
Places and Shapes

Geometry is the study of shapes and space. Young children have considerable experience with geometry before they enter school. They use geometry when they tell where one thing is relative to another ("My coat is on the hook." "The ball is under the chair.") We want to build their vocabulary with position words through songs, games, and activities.

Children explore geometry when playing with two- and three-dimensional shapes. We believe that children should spend time playing, building, and exploring with shapes so that they see relationships among the shapes. Allow children to experiment with how to sort shapes so that you can tell what properties the children know and use, and how they think about shapes (Van De Walle, 2004). Foundational knowledge about shapes and spatial position is critical for learning later math, reading maps, using descriptive language, and recognizing spatial relationships.

The activities in this domain allow children to build the following skills:

- **Demonstrate In & Out**
- **Demonstrate Before & After**
- **Demonstrate Top, Middle & Bottom**
- **Demonstrate Above & Below, Over & Under**
- **Demonstrate Left & Right**
- **Sort Shapes**
- **Match Shapes**
- **Match Shapes of Different Sizes**
- **Move Shapes to Match**
- **Describe Circles**
- **Describe Rectangles**
- **Describe Squares**
- **Describe Triangles**
- **Recognize Shapes in a Group**
- **Identify Shapes in Objects**
- **Explore 3-D Shapes**

Below is the significant research for this domain. For additional Geometry resources, see the reference section at the end of this teacher's guide.

Clements, D.H., and M.T. Battista. 1992. "Geometry and spatial reasoning." *Handbook of Research on Mathematics Teaching and Learning.* Edited by D. A. Grouws. 420-464. New York: Macmillan.

Clements, D.H., and J. Sarama. 2000. "Young Children's Ideas about Geometric Shapes." *Teaching Children Mathematics* 6(8): 482.

Hannibal, M. 1999. "Young Children's Developing Understanding of Geometric Shapes." *Teaching Children Mathematics* 5(6): 353.

4 Squares More Squares®

4 Squares More Squares brings geometry to life in the Pre-K classroom. The chunky rubber pieces are fun for children to handle. Children are engaged to slide, turn, flip, arrange, and rearrange. 4 Squares More Squares captivates Pre-K children with its intriguing shapes and vibrant colors and stretches children's spatial awareness. The product includes:

- 24 Big Pieces — Shapes made up of four squares; each shape has its own color.
- 60 Little Pieces — Single squares; 10 of each color provide variety.
- 12 Boards — Laminated Pattern Boards: rectangles, squares, and steps.
- 6 Colors — Red, yellow, blue, orange, green, and purple make for easy sorting.
- Scoring — Scoring lines on each Big Piece show the four single squares. A scored "X" on one side of each Little Piece distinguishes front and back for easy counting and placement on boards.
- Activity booklet — Activities to explore many areas of math—from puzzles and shapes to counting and patterns.

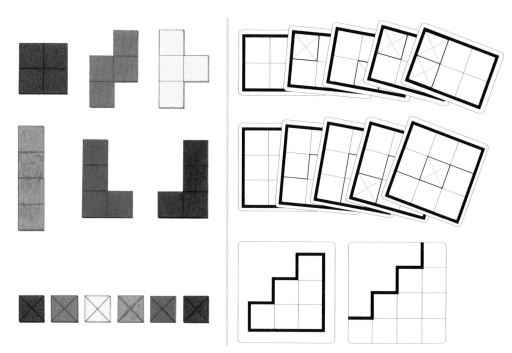

You'll like them because . . .

The product grows with the child as skills advance from counting and matching to pattern making, graphing, and beyond. 4 Squares More Squares invites children to design and build, so they can explore on their own, with friends, or with teacher direction. When children match shapes—move, slide, and flip—and duplicate, create, and grow patterns, they build geometry and algebra foundations.

Where you'll see them

Number & Operations	Measurement & Time
Geometry	Data Representation & Probability
Patterns & Algebra	

Materials
- 2" x 3" Pattern Boards
- 1 of each color of Big Pieces and about 12 Little Pieces

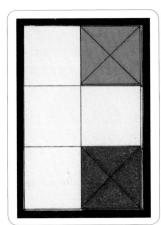

Discovery Play with 4 Squares More Squares®

Delightful teacher-led activities are in the 4 Squares More Squares activity booklet and this teacher's guide. But Discovery Play with 4 Squares More Squares is different. Discovery Play has no set objectives or planned steps. Children are in charge of what they do and what they learn. The learning possibilities are endless, personal, and powerful. We want children to learn beyond what we teach them, to learn what they want to know. As children watch others and become familiar with the materials and teacher-directed activities, their play will change. Ideally, they'll play with more variety and creativity.

Including Discovery Play in Your Day

4 Squares More Squares pieces are versatile and work in all math domains. Use them as counters, measuring units, pattern elements, sorting objects, and more. The activity booklet has a specific section in each activity focused on child's play and exploration. The activities allow children to manipulate the shapes freely as you ask guiding questions. They build oral language skills as they talk about what they are doing with the shapes. They also develop stronger spatial awareness (geometry) and social-emotional skills as they experience concepts through cooperative play. Within any math domain, you can incorporate these manipulatives into daily playtime. Consider an exploration table or area set up with 4 Squares More Squares materials specific to the activity or current theme.

Procedure

Put the boards and the Big and Little Pieces on a table or work/play area. Put the puzzle side up so that children can see where the shapes can fit. Children can explore the different boards and puzzles. They can discover different combination of pieces that fit on the board. Children may turn over the boards to make their own puzzles.

More ideas for Discovery Play:
- Have one partner fill a board and have the other copy their design.
- Use the 3" x 3" Boards or the Steps for children to explore.
- Put out the pieces so children can make their own designs or pictures.
- Children can make color patterns using the Little Pieces.
- Add more pieces or include Mix & Make Shapes™.

Mix & Make Shapes™

Geometry is fun and accessible with Mix & Make Shapes. With multisensory activities, Pre-K children practice problem solving and spatial reasoning as they explore these multicolored shapes. Children discover new ideas with Mix & Make Shapes. The product includes an assortment of shapes.

- Triangles: 4 large, 2 medium, 8 small right triangles, and 8 small isosceles triangles
- Rectangles: 4 large, 4 small
- Squares: 4 large, 4 medium, 8 small
- Circles: 4 large, 4 small
- 4 Semicircles and 8 quarter circles
- 2 Ovals and 2 amoeba shapes

You'll like them because . . .

The large sizes and thick foam pieces are perfect for little hands to place. The colors help children find matching pieces and create their own designs. Children can combine smaller pieces to make large pieces. Each shape is a puzzle waiting to be solved. The versatile shapes are ideal for sorting activities, pattern building, and counting.

Where you'll see them

Number & Operations	**Patterns & Algebra**
Geometry	**Measurement & Time**

Materials
- Variety of Mix & Make Shapes

Discovery Play with Mix & Make Shapes™

Children are naturally drawn to building with Mix & Make shapes. Discovery Play is a time for them to build freely, creating beautiful designs, using shapes to build pictures, and testing how to make tall structures. In this play, they are experimenting with size, area, and patterns while building an understanding of the properties of each shape. Discovery Play will make teacher-led discussions more interesting because children will make connections to their play experiences.

Including Discovery Play in Your Day

Children need time to practice skills that they have learned to generalize them into their world. They should have time to explore every day. Mix & Make Shapes can be a part of this play. Set out specific shapes or specific combination of shapes to encourage curiosity. Children will wonder, "What can I do with the shapes today?"

Procedure - Look What I Made!

- Have shapes available on a table or specific play/work area. Children can work individually or in pairs to create pictures or designs with the shapes. Children will explore in many different ways: making pictures such as a house, face, or flower; making patterns or fitting shapes together.

- Encourage children to talk about their creations. Ask questions about their designs. **What colors did you use? What shape did you put on top? What did you make? How many circles did you use?**

- As children interact with the shapes and with each other, they will use words to describe spatial relationships, position, location, colors, numbers, and shape names.

Other ideas for Discovery Play:

- Make a puzzle: put out the shapes and have children try to match the shapes. You can put out only matching sets or increase the number of shapes for a challenge.

- Sort: put out the shapes and have children sort them by shape, color, or size.

- Add more objects: include some of the pieces or Pattern Boards from 4 Squares More Squares®.

In or Out? Toss & Shout!

Demonstrate In & Out

Position words are important for language development and are a key component of geometry. When children know position words, they can describe the relationship between objects, such as in and out. Use position words regularly so they become part of children's vocabularies.

Look What We're Learning

Geometry
- Identify position or location using in and out

Sensory Motor
- Naturally move and place body to perform tasks

Social-Emotional
- Take turns

Vocabulary

in

out

In or Out? Toss & Shout!

Materials/Setup:
- Hula hoop (or mark an area about that size)
- 1 Tag Bag® of each color

Grouping:
Small group (6)

English Language Learners:
Name the colors and hold up the corresponding Tag Bag before the activity. After each color, have children repeat the color names aloud. Children will be able to participate by watching peers. Tossing will be easy.

Objective
Children describe position using in and out.

Activity
Will we toss our Tag Bags in or out? Let's see!

1. Place Tag Bags inside the hula hoop on the floor. **Take one Tag Bag.**

2. **Take five big steps out from the circle.** Show which direction is out.

3. Have children toss by color. **Who has red? Take one big step in. Toss and shout, IN or OUT?**

4. Repeat for other colors.

✓ Check for Understanding

Observe children as they toss the Tag Bags. Do they shout the correct word to describe whether Tag Bags are in or out?

Support: Have children place the green Tag Bag in the hoop and the yellow Tag Bag out of the hoop. Help children use in and out to talk about the position of each Tag Bag.

More to Learn

Cups and Caps
Ask children to place a bottle cap in a cup. Take turns asking the child to place a cap in or out of the cup and asking the child to tell where he placed it. Use the cups and caps activities in *I Know My Numbers* to practice in and out (use booklets 1–3, 6, and 7).

Count It Out
Challenge children to count the Tag Bags that are out. Count the Tag Bags that are in. Count how many Tag Bags in all.

Let's Make a Rainbow!

Demonstrate Before & After

Children know before and after in relation to ordering events in time, "I take off my clothes before I take my bath." Geometry is all about how things are positioned in space. Here they will learn to use before and after in reference to things in a line.

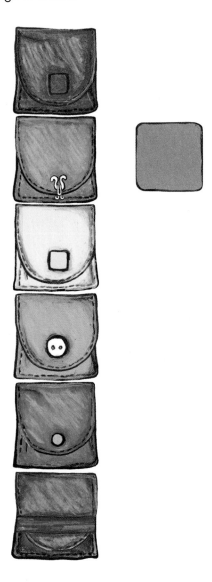

Look What We're Learning

Geometry
- Identify position or location using before and after

Social-Emotional
- Take turns

Literacy
- Listen to follow directions

Vocabulary

before

after

Let's Make a Rainbow!

Materials/Setup:
- 1 Tag Bag® of each color
- 1 Color Tag of each color

Grouping:
Small group (6)

English Language Learners:
Show a picture of a rainbow at the beginning of the activity. Name each color on the rainbow and have children repeat. Invite children to name colors in their home language.

Objective
Children describe position using before and after.

Activity
Let's play a game using before and after.

1. **Have you ever seen a rainbow? Tell me about it.**

2. Give six children each a Tag Bag of a different color. **Let's put the Tag Bags in rainbow order. Who has red? We will start here.**

3. **Who has orange? Stand after red.** Continue with other colors. Make sure children are facing the same direction. **You are the rainbow.**

4. Keep children in rainbow standing in line. **Let's play! I'm going to pick a Color Tag without looking. I picked yellow. What color is before yellow? Orange.**

5. Invite a child to pick a Color Tag. Help find the Tag Bag that comes before the color on the Color Tag.

6. Play again, with a different child picking a Color Tag. This time pick the Tag Bag that comes after the color you draw. You can also give children different color Tag Bags and line up another time.

✓ Check for Understanding

Observe children as they arrange the Tag Bags. Do they understand the concept of before and after?

Support: Practice with just three colored cubes: red, orange, and yellow. Introduce the red, orange, yellow left-to-right sequence. Place red. Touch and say, **Red.** Place orange. Touch and say, **Orange is after red.** Place yellow. Touch and say, **Yellow is after orange.**

More to Learn

Who's Before?
When children get into line, talk about where they are in relation to friends. **Tyrone is before Najib. Najib is after Tyrone.**

What's on First?
Connect this activity to ordinal numbers. Help children understand each ordinal number and what is before and/or after it. Have children use before and after and the ordinals to describe the Tag Bags in line.

Red Light, Green Light

Demonstrate Top, Middle & Bottom

It is important that children understand top, middle, and bottom so they can describe position in geometry. They also need to understand directions. We use these terms when teaching children how to form their letters and numbers. We also teach them to read from top to bottom.

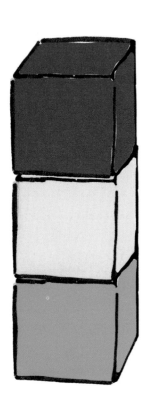

TRAFFIC LIGHT

© 2008 Jan Z. Olsen

Look What We're Learning

Geometry
- Identify position or location using top, middle, and bottom
- Draw a picture to help solve a problem

Problem Solving
- Draw a picture to help solve a problem

Social-Emotional
- Take turns

Literacy
- Listen to follow directions

Sensory Motor
- Move an object in one hand to position it for release
- Look at hands and use visual cues to guide reaching for, grasping, and moving objects

Vocabulary

bottom

middle

top

tower

Red Light, Green Light

Materials/Setup:

- Red, yellow, and green cubes (1 per child)
- 4 Squares More Squares® red, yellow and green Little Pieces
- Red, yellow, and green crayons
- Traffic Light handout from A Click Away (1 per child)

Grouping:

Small group (10)

English Language Learners:

Hang the stoplight picture before the activity. Say the word and have children repeat it. **Have you seen this before? Where?** Point to the top. **What color is at the top?** Repeat steps for middle and bottom.

Objective

Children use top, middle, and bottom to describe position.

Activity

Let's learn about top, middle, and bottom.

1. Build a cube tower. **Green is at the bottom. Yellow is in the middle. Red is at the top. What does this look like? A traffic light!**

2. Have pairs of children make towers that match. Invite them to describe their towers.

3. Give each child red, yellow, and green Little Pieces. **Let's build a traffic light. Put the red at the top.** Continue with the middle and bottom.

4. For more practice, invite children to color the Traffic Light page. **Let's color red on top.** Continue with middle and bottom.

✓ Check for Understanding

Ask students to build a tower of blocks with yellow on bottom, red in the middle, and green on top. Do not provide a model. Do their towers match your directions?

Support: Sing "Where Do You Start Your Letters?" from the *Get Set for School Sing Along* CD, track 1. Dance and show children where to find the top, middle, and bottom on their own bodies.

More to Learn

I Found the Top

Build a picture with your 4 Squares More Squares pieces. Use top, middle, and bottom to explain where pieces should go. See "Look, I Made a Flower!" (second activity) in the 4 Squares More Squares activity booklet.

What's in the Middle?

Extend the lesson by asking children to build a tower of four cubes. It's easy to point to the top and bottom, but more challenging to understand that two blocks are in the middle.

Listen & Pass

Demonstrate Above & Below, Over & Under

Emphasize the words above and below throughout the day to build children's knowledge of position words. They will learn with the repetition and will soon be able to describe the positions of things accordingly.

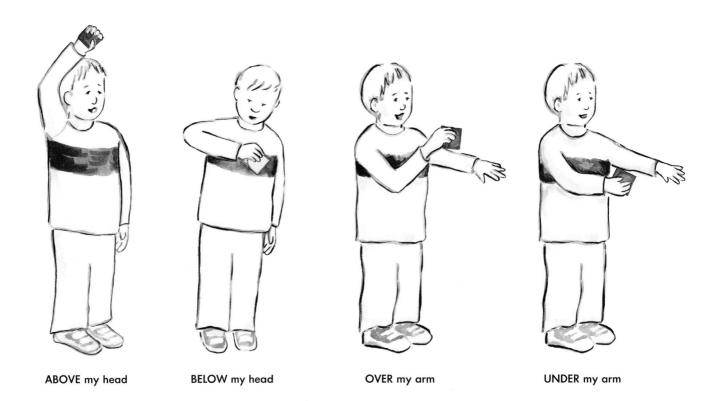

ABOVE my head   BELOW my head   OVER my arm   UNDER my arm

Look What We're Learning

Geometry
- Identify position or location using above and below, over and under

Literacy
- Listen to follow directions

Sensory Motor
- Move an object in one hand to position it for placement or release

Vocabulary

above

below

over

under

Materials/Setup:

• Tag Bags®

Grouping:

Small group or whole class

English Language Learners:

Encourage children to say the words as they pass. Have them say **above** with each pass above their heads.

Objective

Children describe position using above and below, over and under.

Activity

We are going to pass the Tag Bag. Listen carefully to my instructions.

1. **Let's pass the Tag Bag above our heads.** Pass around the circle once. Have children say, "The Tag Bag is above my head" as they pass the Tag Bag.

2. **Now let's pass the Tag Bag below our head.** Pass around the circle once. Have children say, "The Tag Bag is below my head" as they pass the Tag Bag.

3. **Listen carefully to find out how to pass.** Tell children whether to pass it above or below. Don't use a pattern so that they have to listen to the words.

4. Repeat the activity. This time, show how to pass **over** your arm. Pass around. Then show how to pass **under** your arm. Pass around. Finish with a listening round using over and under.

✓ Check for Understanding

Observe children as they respond to your directions. Do they follow your directions when they pass the Tag Bag?

Support: Bring prepositions into the building block area. Use a person or animal figurine in a child's block structure. Tell whether it is above, below, over, or under the structure.

Change the movements to accommodate students with limited mobility.

More to Learn

Tower Power

Have children build a tower with four different color blocks. Point to one block. **Green is under purple. What is under green?** Repeat using above, below, and over.

Where Does It Go?

For a challenge, take a walk around the room and find things that are above, below, over, and under. **The chair is under the table. The clock is over the door.**

Shake Hands With Me

Demonstrate Left & Right

It takes time to distinguish between left and right. The key is to teach only the right hand. Different sensory experiences help children learn the right hand. When they know the right, they know the left. An easy way to do this is to teach children to always shake hands with the right hand.

Look What We're Learning

Geometry
- Identify position or location using left and right

Social-Emotional
- Take turns

Sensory Motor
- Look at hands and use visual cues to guide reaching for, grasping, and moving objects

Vocabulary

left

right

Shake Hands With Me

Materials/Setup:
- *Get Set for School Sing Along* CD track 7, "Hello Song"
- May use one or none:
  - Lotion
  - Rubber stamp
  - Water
  - Sticker

Grouping:
Small group or Whole class

English Language Learners:
Let children watch others first. Continue the activity with other multisensory experiences. Children can take cues from their classmates to participate fully in the activity.

Objective
Children describe position using left and right.

Activity
Let's learn to use the right hand to shake hands.

1. Shake hands with each child. Smile and make eye contact. **Hello.**

2. Give one sensory experience to each child's right hand. **This is your right hand. I'm going to do something fun to your right hand.**

   - Rub right hand

   - Dab lotion on right thumb and index finger for child to rub

   - Dip child's right fingertips in water

   - Put sticker on child's right hand

3. Tell children to raise their right hands and repeat after you. **This is my right hand.** "This is my right hand." **I shake hands with my right hand.** "I shake hands with my right hand."

4. Play "Hello Song." Have children shake hands while singing the song.

✓ Check for Understanding

Watch children during the song. Do they automatically put out their right hands when you approach to shake hands?

Support: Use care with children who are sensitive to touch. Give them a stimulus that is agreeable to them.

More to Learn

Learn Left
When children can always show you the right hand when asked, follow by asking about the left. **Which hand is left?**

Left Hand, Right Hand
Use *I Know My Numbers* booklet 2 to practice left and right. Use cups and caps. Place a cap on the table. **Place a cup on the right side of the cap.**

Sides & Corners

Sort Shapes

Children are more likely to remember information that they discover through experience. They need to handle shapes and straight sides, corners, and curves to understand the characteristics of shape. They can organize or sort the shapes by these characteristics to prepare for describing shapes.

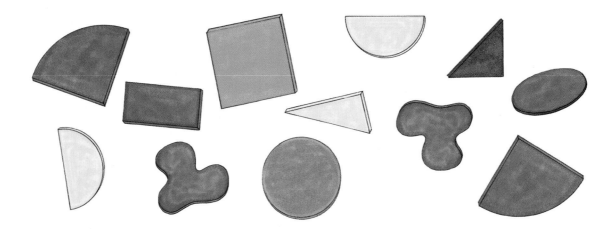

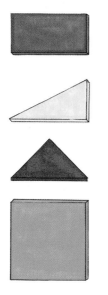  

Look What We're Learning

Patterns & Algebra
- Sort objects by shape
- Identify objects or pictures as the same or different

Number & Operations
- Verbally count a set of five objects

Social-Emotional
- Take turns
- Work with others to solve problems

Vocabulary

corner
side
straight

Sides & Corners

Materials/Setup:
- Mix & Make Shapes™:
  - 2 Quarter circles
  - 2 Half circles
  - 1 Square
  - 2 Different triangles
  - 1 Rectangle
  - 1 Circle
  - 1 Oval
  - 2 Amoeba shapes

Grouping:
Small group (3)

English Language Learners:
Point to sides and corners when you talk about triangles and rectangles. Use the word **round** and run your finger around the circle. Let the child feel the shapes.

Objective
Children sort shapes by sides and corners.

Activity
Who knows about sorting? Today we are going to sort shapes.

1. Spread the shapes on the floor in the middle of the group. Ask one child to pick a shape.

2. **You picked a great shape, Angie! It has straight sides and corners.** Point to the sides and corners as you name them.

3. **Let's see if we can find more shapes like Angie's. They must have straight sides and corners.** Allow children to find other shapes. Help if they get stuck. Gently remind about the sorting rule if a shape with a curved side is chosen.

4. **Good job! Let's see how many shapes are in Angie's pile. 1, 2, 3, 4. There are 4 shapes with straight sides and corners.**

5. Have another child pick a shape. Repeat the activity sorting out:
   - Shapes with curved sides and no corners
   - Shapes with curved sides, straight sides, and corners

6. Each of the children should have a pile of shapes at the end. Ask them to describe their shapes.

✓ Check for Understanding
Ask children to describe their pile of shapes. Do they talk about sides and corners correctly?

Support: Use 12 shapes. Sort only shapes with lines and corners (squares, triangles, rectangles) and shapes with only curves (circles, ovals, amoeba shapes).

More to Learn

How Many Can You Find?
Use *I Know My Numbers* to practice counting sides and corners (use booklets 4 and 8).

Sort Some More!
Ask children to think of another way to sort the shapes. Help them talk about their rules for sorting.

Guess What?

Match Shapes

Impressions are a great way for children to match shapes. When they match objects to impressions, they notice characteristics of both. The real object has an outline, shape, size, and texture that show in the impression. Talk with children about such characteristics to boost vocabulary and thinking. Use this activity to talk casually about shape characteristics.

Look What We're Learning

Geometry
- Match shapes of same size, shape, and orientation

Patterns & Algebra
- Identify objects or pictures as same or different

Problem Solving
- Use manipulatives to find a solution
- Guess and check the answer; repeat until correct answer is found

Social-Emotional
- Take turns

Vocabulary

check

match

round

Guess What?

Materials/Setup:
- Small tray (per child)
- Dough
- Blocks, buttons, caps, forks, rings, keys, beads

Grouping:
One-on-one; pairs

English Language Learners:
Ask child to repeat the names of each object in the group. When you talk about the characteristics of a shape, point and show what you mean.

Objective
Children match objects to impressions by noticing shape, size, and orientation.

Activity
Choose one thing to press in the dough. I will guess what you used.

1. Have children spread the dough inside their trays.

2. Show the objects. Let children name the objects. Name any items they don't know.

3. **While my eyes are closed, press one thing into your dough. Don't let me see. Put it back in the pile.** Child may need help. **Say "open" when you are done.**

4. Open your eyes. Describe the characteristics of the impression. **Your shape is round. It has two bumps in the middle. I think you used a button. Let me check.**

5. Test button on your dough. **Yes, you used a button. This matches! It's the same!**

6. **My turn, close your eyes.** Press one thing into the dough. Have children describe what you pressed. Have them check.

✓ Check for Understanding

Is the child able to tell whether they are the same or different? Do they name the object and begin to describe the impression?

Support: Play the game with eyes open first. Make an impression in your dough and have the child repeat after you. Take turns. When you play with eyes closed, have only three or four objects in the group.

More to Learn

Match My Shape
Hand out small Mix & Make Shapes™ to a group of children (use circles, triangles, and rectangles). Let them find their shape family.

A Funny Face
Make the activity more challenging by stamping more than one shape at a time. Stamp out a funny face and have a child copy it.

Are You My Match?

Match Shapes of Different Sizes

Shapes can be similar even when they are different sizes. A shape with three sides and three corners is a triangle no matter what size it is. Your children are ready to learn this by matching shapes of different sizes. This activity helps them learn the characteristics to notice (sides and corners) and the characteristics to ignore (size and color) when they sort by shape.

Look What We're Learning

Geometry
- Recognize that shapes can be the same even if sizes differ

Number & Operations
- Verbally count a set of five objects

Social-Emotional
- Take turns

Vocabulary

big

corner

little

same

shape

Are You My Match?

Materials/Setup:

- Mix & Make Shapes™:
  - Circles – 1 large, a few small*
  - Triangles – 1 large, a few small*
  - Rectangles – 1 large, a few small*

  *each child needs one small shape

Grouping:

Small group or whole class

English Language Learners:

Practice the pattern of question and answer with this activity. Model how to rephrase the question into an answer. **Are you my shape? No, you are not my shape.**

Objective

Children find similar shapes of different sizes.

Activity

The little shapes want to find something big that is the same shape. Can you help?

1. Ask three students to stand in a line. Give each child a big shape.

2. Hand out small shapes to the other children in the group. Keep a rectangle for yourself.

3. Show how to check for a match. **Are you my shape? No! My shape has four corners. Your shape doesn't have corners.** Continue until you find the big rectangle.

4. Give each child a chance to find a big shape of the same kind. Encourage them to talk about why their shapes are the same or different.

✓ Check for Understanding

Post the big shapes in different corners of the room. Give children a small shape. Ask them to move to the big shape that matches their shape. Are they able to make a match?

Support: Start with the circle and triangle. Talk about corners and sides to explain why shapes match.

More to Learn

My Shapes Match

Use the 4 Squares More Squares® pieces to match different size shapes. Use the red, green, and blue Big Pieces. Give a yellow Little Piece to each child and have her find the big shape that matches (red).

I Like Squares, Too

For an extra challenge, include big and small squares in the activity. Talk about what the squares and rectangles have in common.

Puzzle Moves

Move Shapes to Match

Shapes can be hard to match when they are oriented differently. A simple turn or flip reveals the match. Turns rotate the piece into place. Flips turn the shape over to show its mirror image. Puzzles are a good way to explore orientation. Children move pieces to match the shape left on the puzzle.

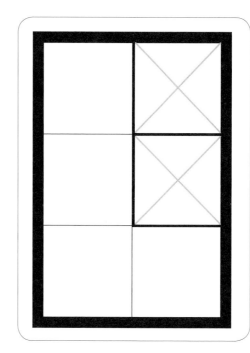

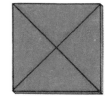

Look What We're Learning

Geometry
- Recognize that shapes can be the same even if positioned differently

Number & Operations
- Verbally count a set of five objects

Social-Emotional
- Work with others to solve problems

Vocabulary

fit

flip

turn

Materials/Setup:

- 4 Squares More Squares®:
  - Big and Little Pieces in all colors
  - 2" x 3" Pattern Boards (1 per pair)

Grouping:
Pairs

English Language Learners:

Practice puzzle moves on the carpet before the activity. Demonstrate slide (sit and scoot forward), turn (sit and spin), and flip (lie flat on back and roll over). Have children join you.

Objective
Children turn and flip shapes to build a puzzle.

Activity
Let's build a puzzle. You'll find the Big Piece that goes on your puzzle.

1. Place a 2" x 3" Pattern Board in front of each pair with the puzzle side up. Assign one partner to Little Pieces and one to Big Pieces.

2. **Look at your board. The shapes are outlined. Put two Little Pieces on the puzzle.**

3. **Look at the space that is left. Which Big Piece will fit on your puzzle?** Have the other partner choose a shape and try it on the board. If a pair of children is struggling to make a shape fit, suggest a turn or flip and demonstrate.

4. Have children tell you about their puzzle.
   How did you move your Big Piece to make it fit? How many Little Pieces did you use? Big Pieces?

5. Give each pair a new board. Have partners switch roles.

✓ Check for Understanding

Observe children as they build the puzzle. Can they choose the piece that will fit? Can they do this even if the piece is in a different orientation than the space on the puzzle?

Support: Use only the red and yellow Big Pieces and their corresponding puzzles.

More to Learn

Match My Shape
Scatter the big shapes on the carpet. They will naturally fall in different orientations. Give children a shape and ask them to stack the shapes that match.

Puzzle Challenge
Challenge children by using the plain side of the Pattern Board. One child picks a Big Piece, then the other completes the puzzle with Little Pieces. Give them paper copies of the board to color their solutions. See how many solutions they can find.

Round & Round We Go

Describe Circles

Circles are the easiest shape for children to recognize. They can be found all around—in wheels, buttons, and pancakes. Circles make the letter O and number 0 (zero). With this activity, children describe what makes this familiar shape a circle.

Look What We're Learning

Geometry
- Identify and describe circles

Number & Operations
- Verbally count a set of five objects

Problem Solving
- Use manipulatives to find a solution

Social-Emotional
- Take turns

Vocabulary

corner

curve

roll

round

side

Round & Round We Go

Materials/Setup:
- Mix & Make Shapes™:
  - 2 Large circles
  - 2 Small circles
  - 4 Half circles
  - 8 Quarter circles

Grouping:
Small group (4)

English Language Learners:
Use this opportunity to build vocabulary. Help children find and name things that are circles (e.g., wheel, clock, cracker). Children may name objects in their home language. This is okay. Encourage them to speak in any language so that they feel comfortable and part of the class.

Objective
Children identify and describe circles.

Activity
Today we will build a shape that goes round and round.

1. Show a big circle. **Circles have one round side and no corners. They can roll smoothly along the floor.**

2. Give each child a circle. **Roll your circle with me. Roll it on your arm. Roll it over your head. Roll it down your body.**

3. Pair students who have green circles with students who have orange circles. Give each child a yellow half circle.

4. Put your orange circle on the ground.
   I have a trick, a trick for you.
   I can make one out of two.
   Put two yellow half circles on top of the orange circle.
   Now you try!

5. Give each child two blue quarter circles. Put your green circle on the ground.
   I have a trick, a trick on the floor.
   I can make one out of four.
   Put four blue quarter circles on top of the green circle.
   Now you try!

✓ Check for Understanding

After children build circles ask questions. **Did you build a circle? How do you know?** Are they able to describe circles?

Support: Have children trace the small orange circles. Ask them to tell you about their circles.

More to Learn

Curious Curves
You can also use Wood Pieces for Capital Letters. Use Big Curves and Little Curves to build circles.

Is It a Circle?
For a real challenge, show students all the pieces with curved sides. **Which ones are circles? How do you know?**

Our Rectangle Book

Describe Rectangles

Rectangles are easier to draw than circles—they are in doors, windows, boxes, and books. Rectangles have four sides: the opposite sides are parallel and the same size. Rectangles also have four right angles (square corners). Teach these properties so children can distinguish rectangles from other quadrilaterals (shapes with four sides and four angles).

Look What We're Learning

Geometry
- Identify and describe rectangles

Sensory Motor
- Move object in one hand to position for use
- Look at hands and use visual cues to guide moving objects

Vocabulary

corner

opposite

same

side

Our Rectangle Book

Materials/Setup:

- Mix & Make Shapes™:
  - 1 Large rectangle
- Rectangular sponges of many sizes (1 per child)
- Washable paint
- Paint tray or paper plate (1 per color)
- Blank rectangular paper
- Smocks (1 per child)

Grouping:

Small group

English Language Learners:

Have children say the word for rectangle in their home language. If they can relate the word to vocabulary they already know, they will deepen their understanding of the concept.

Objective

Children identify and describe rectangles by creating a book.

Activity

We are going to make a shape book about rectangles.

1. Show the big green rectangle. Point as you describe. **This is a rectangle. It has four sides. The opposite sides are the same. It also has four corners. They are all the same.**

2. Show a sponge. **This sponge has many rectangles. There's one on each side. Let's use our sponges to make rectangles for our book.**

3. Show how to make a rectangle by dipping sponge in paint. Demonstrate vertical, horizontal, and diagonal rectangles.

4. Invite children to paint rectangles using the sponges.

5. When paint is dry, ask children to describe painting (e.g., shape, number, size, orientation, color). Write their exact words on the page. Join the pages to create a rectangle book.

✓ Check for Understanding

Have children read the rectangle book to you. Can they identify rectangles? Can they describe the sides and corners?

Support: Cut Post-it® notes vertically to make rectangles (so each side has a sticky part). Take children on a rectangle tour in the classroom or hallway. Stick Post-its saying, **Rectangle on a rectangle, rectangle on a rectangle.**

More to Learn

Count How Many
Use the 4 Squares More Squares® 2" x 3" Pattern Boards to build rectangles. Count how many different combinations you can use to make a 2" x 3" rectangle.

Stop and Turn
To challenge students, have them trace rectangles. Encourage them to describe as they trace. **Long line, stop and turn, short line, stop and turn . . .**

1 Purple, 2 Purple, 3 Purple, 4

Describe Squares

The square can be a tricky shape for children. It is a special rectangle—all four sides are the same. When squares are rotated, children may call them diamonds. The difference between squares and diamonds is the angles. Squares have four angles that are all the same. In diamonds, the opposite angles are the same. Teach children to look for sides and corners that are all the same so that they can spot squares in any orientation.

 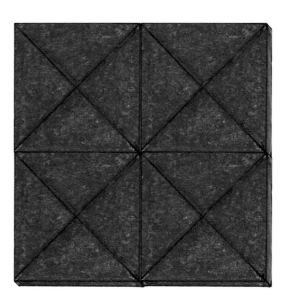

Look What We're Learning

Geometry
- Identify and describe squares

Number & Operations
- Verbally count a set of five objects

Problem Solving
- Use manipulatives to find a solution

Social-Emotional
- Work together to solve a problem

Vocabulary

corner

same

side

1 Purple, 2 Purple, 3 Purple, 4

Materials/Setup:
- 4 Squares More Squares® Little Pieces
- 1 Red Big Piece

Grouping:
Whole class (multiples of 4)

English Language Learners:
Children can say the chant as they pass out the squares as they are able, depending on their language level. Give an alternative of just saying or singing the numbers. When asking for volunteers, don't put children on the spot— guide them to participate. Some will be more eager than others.

Objective
Children identify and describe squares.

Activity
Today we will build a shape that has sides and corners all the same size.

1. Hold up a red Big Piece. **This is a square. It has four sides that are all the same. It also has four corners that are all the same.** Hold up a Little Piece. **This is a little square. It has four sides and four corners all the same.**

2. Give away four Little Pieces of each color by chanting a version of "One Potato, two potato." Repeat the chant for each color:
**One purple, two purple, three purple, four
Four purple people build a square on the floor**

3. Show each color group where to build. **Put your little squares together to make a big square.**

4. **Are all the sides the same?** Show that each side is two little squares long.

5. **Are all the corners the same?** Show that each corner is the same by measuring with another Little Piece.

✓ Check for Understanding

Ask children to tell you about their squares. **How do you know it is a square?** Are they able to identify squares?

Support: Provide a red Big Piece to the group of children as a template for making a big square. Children can stack the pieces on top to make a square.

More to Learn

Building a Big Square
For a real challenge, children can make a class square using 4 Squares More Squares Little Pieces. Have each group add the big square to the class square. This works with four or nine groups.

Cover the Square
Search for other ways to build a square using Mix & Make Shapes™. Use a green square as template. Cover with four red squares, four red triangles, two purple rectangles, or four yellow triangles.

Terrific Triangles

Describe Triangles

Builders know that a triangle is one of the strongest shapes. It is used in bridges, roofs, and trusses. Teach children that this strong shape has three sides and three angles (corners). Triangles come in many different varieties. Show children examples of each.

Look What We're Learning

Geometry
- Identify and describe triangles

Number & Operations
- Verbally count a set of five objects
- Make or draw a set of objects to match a given number

Sensory Motor
- Look at hands and use visual cues to guide moving objects

Vocabulary

corner

side

Terrific Triangles

Materials/Setup:

- Mix & Make Shapes™:
  - 1 Large triangle
  - 1 Medium triangle
- Wood Pieces:
  - Big Lines (3 per child)
  - Little Lines (3 per child)

Grouping:

Small group or whole class

English Language Learners:

During the activity, as you are showing the triangle, point to the sides and corners as you name them. Children can repeat the names. Take your time and even repeat the names. Count the corners and sides aloud as a class. These may be new words for everyone.

Objective

Children build and describe triangles.

Activity

Let's build a shape with three sides and three corners!

1. Show the big purple triangle. **This is a triangle. It has three sides and three corners. Where have you seen this shape?**

2. Have three children lay down on the rug to make a triangle. Give any volunteers a chance to try.

3. Show children how to build a triangle with the Big Lines. Let them build.

4. Give children two Big Lines and one Little Line. Ask them to build another triangle. **Is this a triangle? How do you know?**

✓ Check for Understanding

Encourage children to use the Big Lines and Little Lines to make more triangles. Can they build a triangle? Do they have the right number of sides and corners?

Support: Give each child three caps. Let children place the caps on each corner of a triangle. Have them count while placing caps.

More to Learn

Sing and Draw

Have children draw the shapes. Use "My Teacher Draws" from *Get Set for School Sing Along* CD to show them how.

I Can Make a Triangle

Build a triangle with Mix & Make Shapes. Use a purple triangle as a template. Cover with two green triangles or eight yellow triangles. For a real challenge, build with one green square and four yellow triangles.

Simon Says Shapes

Recognize Shapes in a Group

When children can name shapes in isolation, they are ready to name shapes within a group. This skill requires them to recall the name and characteristic of each shape. Children must also sort that information in their minds when they look at all the shapes.

Look What We're Learning

Geometry
• Identify specific shapes within a group

Social-Emotional
• Participate in clean-up routines
• Cooperate with other children

Simon Says Shapes

Materials/Setup:
- Mix & Make Shapes™:
  - 4 Large circles
  - 4 Large triangles
  - 4 Large rectangles
  - 4 Large squares

Grouping:
Small group (4)

English Language Learners:
Preview the actions before starting the activity. Model and say the name of the movement: jump, hop, hold up your hand, tickle, sit. Children can repeat the name and imitate your movements.

Objective
Children recognize circles, triangles, rectangles, and squares within a group.

Activity
Let's play Simon Says. Simon will tell you what to do with the shapes.

1. Spread the shapes on the ground. **I will be Simon first. I'll tell you what to do.**

2. Give directions. **Simon says:**
   - **Jump to a triangle.**
   - **Put your hand on a square.**
   - **Tickle a circle.**
   - **Sit on a shape with four sides.**

3. When they understand the activity, let a child be Simon. Play along and help if the group gets stuck.

4. Don't try to trick the children until they can play without mistakes. Then give a direction without saying **Simon says.**

✓ Check for Understanding

Have children clean up the pieces. Ask each to pick up a different shape. Do they select the right shapes?

Support: Start with two shapes. Add additional shapes when children consistently recognize the first two.

More to Learn

Body Shapes
Read *Mat Man Shapes*. Build Mat Man® with different shapes for his body. Use the patterns at the back of the Mat Man book.

More Simon Says
Add more shapes to Simon Says for an extra challenge. You can also use an oval, star, diamond, pentagon, hexagon, or octagon.

Snap a Shape

Identify Shapes in Objects

Shapes are all around us. The ability to identify shapes helps children make sense of their environment. This is a first step for aspiring artists, engineers, architects, and city planners.

You know how important it is to prepare children for any learning experience. Read a book that shows shapes in real world objects. This will help them know what to do and keep them focused when they go for a walk to look for shapes.

Look What We're Learning

Geometry
- Identify shapes in real world objects
- Identify specific shapes within a group

Patterns & Algebra
- Sort objects by shape

Social-Emotional
- Manage and handle transitions well and without incident
- Treat property with respect

Materials/Setup:

- Camera
- Shot list

Grouping:
Small group (4)

English Language Learners:
Use the book to review names of shapes and objects. Point to the picture and name the object. During the walk, help children name the objects they choose to photograph.

Objective
Children recognize and name shapes in real-world objects.

Activity
Shapes are everywhere! Let's hunt for shapes in the world around us.

1. Read a book about shapes in real-world objects. **Which shapes do you see on each page?** Invite children to point out shapes in the book.

2. **Do you think there are shapes in our school? Let's go on a walk. We can snap pictures of any shapes we see.** Take a walk and encourage children to share any shapes they find. Take pictures. Use the shot list to remember who found each shape.

3. Print the photos.

4. Invite children to sort the photos by shape. Display in the classroom.

✓ Check for Understanding

Observe children as they identify shapes in their environment. Does each child find a shape during the walk? Can children name the shapes? Can they sort by shape?

Support: Place a transparency over the pages of the book. Outline a shape with an erasable marker. Invite children to do the same.

More to Learn

Real World Shapes
Use *My First School Book* activity book pages 10–19 to discover and color shapes in real world objects.

My Book About Shapes
Extend this activity by creating shape books. Make a book for each shape with children's pictures. Let the photographers tell you about their pictures. Write their words under the pictures. If you have a class blog, this is a great place to post pictures and captions.

Stack It Up

Explore 3-D Shapes

Pre-K children are very familiar with three-dimensional shapes from the block area. By 18 months, they can build a tower of up to four blocks. Experience has shown them that flat objects can stack and round ones cannot. Use this activity to help them talk about this knowledge and organize their thinking.

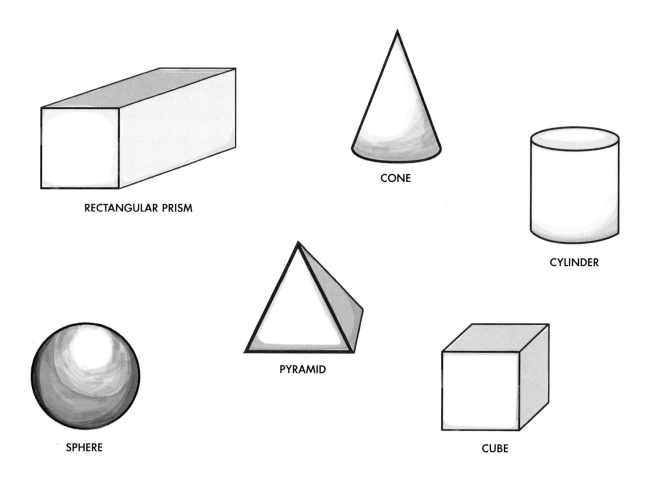

RECTANGULAR PRISM

CONE

CYLINDER

PYRAMID

SPHERE

CUBE

Look What We're Learning

Geometry
- Recognize three-dimensional shapes

Patterns & Algebra
- Sort objects by function/kind

Problem Solving
- Use manipulatives to find solutions

Vocabulary

stack

Stack It Up

Materials/Setup:

- Multiple blocks:*
  - Spheres
  - Cubes
  - Rectangular Prisms
  - Cylinders
  - Pyramids
  - Cones

  *Use recycled food containers if blocks aren't available

Grouping:

Small group

English Language Learners:

Invite a child to stack with a set of cubes. **You stacked the blocks.** Point to the tower. **This is a stack.** Encourage the child to tell you what he did.

Objective

Children describe and sort three-dimensional shapes.

Activity

Let's find out which blocks can stack.

1. Lay out blocks in the work area.

2. Select a helper. Ask the helper to stack the cubes. **Do the cubes stack?** (Yes.) **Do they stack if you turn them?** (Yes.)

3. Choose another helper. Ask the helper to stack the spheres. **Do the spheres stack?** (No.) **Why won't they stack?** (The sides are round, so they roll.)

4. Have children make two piles, shapes that stack and shapes that don't stack.

5. Repeat with the other blocks.

✓ Check for Understanding

Observe children as they stack the three-dimensional shapes. Do they understand which blocks can stack and which ones can't? Can they stack with the prism and cylinder?

Support: Work together to stack the blocks. Have child add first block, then you add second block. For cylinder, try stacking round surfaces first. Let child correct you.

More to Learn

Hunting for Shapes
Find shapes in the classroom that match the three-dimensional shapes.

Shapes on My Shapes
Ask children to point out the two-dimensional shapes they see on the faces of each three-dimensional shape.

PATTERNS & ALGEBRA

See It, Repeat It

Children use algebra when they start to sort. Even as toddlers, they may sort their food, their toys, or their clothes. Sorting and classifying objects is a fundamental skill and the basis for understanding many concepts in both science and math. In our activities, children use specific characteristics to sort objects. They learn that the same objects can be sorted in different ways, such as by color or by shape.

The identification and extension of patterns is also an important process in algebraic thinking. Children love to notice and make patterns. Seeing and extending simple patterns promotes observation, thinking, and problem-solving skills. Pattern activities using manipulatives allow them to extend a pattern and make changes without fear of being wrong (Van De Walle, 2004). We use sound, movement, color, and shape patterns to help children develop problem-solving strategies.

The activities in this domain allow children to build the following skills:

- **Classify Same or Different**
- **Sort by Color**
- **Sort by Size**
- **Sort by Function**
- **Describe a Simple Pattern**
- **Duplicate a Simple Pattern**
- **Explore Growing Patterns**
- **Explore Patterns in the Real World**

Below is the significant research for this domain. For additional Patterns & Algebra resources, see the reference section at the end of this teacher's guide.

National Association for the Education of Young Children and the National Council of Teachers of Mathematics. 2003. Joint position statement: *Early Childhood Mathematics: Promoting Good Beginnings.* Washington, D.C.

Taylor-Cox, J. 2003. "Algebra in the Early Years? Yes!" *Young Children.* 58(1): 14-21.

Wardle, F. 2007. "Math in Early Childhood." *Exchange.* 55-58.

One Is Different

Classify Same or Different

Before children can sort multiple objects into groups, they need to be able to tell if the objects are the same or different. Even as babies, they already know same and different. What's new is using words to describe objects and then saying why they are the same or different. They learn a process for describing and deciding.

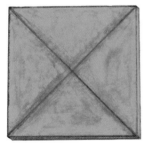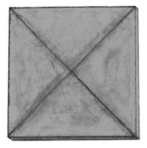

Look What We're Learning

Patterns & Algebra
- Identify objects or pictures as the same or different

Literacy
- Ask and answer simple questions

Vocabulary

different

same

One Is Different

Materials/Setup:
- 4 Squares More Squares®:
  – 1 Purple Little Piece
  – 3 Green Little Pieces
- Coins or play money:
  – 3 Pennies
  – 1 Nickel

Grouping:
Small group

English Language Learners:
Give children the objects to touch and explore. They will discover much more than they would just by looking. When possible, use manipulatives and let children play with them in their own way. Manipulatives make concepts real and tangible.

Objective
Children describe same and different objects.

Activity
Let's look at a few things. You tell me what's the same and different.
1. Show three green Little Pieces and one purple Little Piece.
2. Talk about same. **How are the squares the same?** Help children describe how they are the same.
3. Talk about different. **How is one square different?** (Color.)
4. Repeat the activity with three pennies and one nickel.

✓ Check for Understanding

Observe as children use adjectives to describe size, shape, and color. Listen for explanations about same or different decisions. Do children use the vocabulary in their explanations?

Support: Show children two blue squares. **Are they the same color/size/shape?** Help children see that they are the same. Now show one blue and one orange square. **Are they the same color?** Help children see that they are different.

More to Learn

More Money
Repeat the coin activity with pennies, nickels, and dimes.

Different Sizes
Repeat the square activity with one red Big Piece.

Color Sort

Sort by Color

Children prepare for sorting when they talk about same and different. Sorting by color reinforces color names and focuses a child's attention.

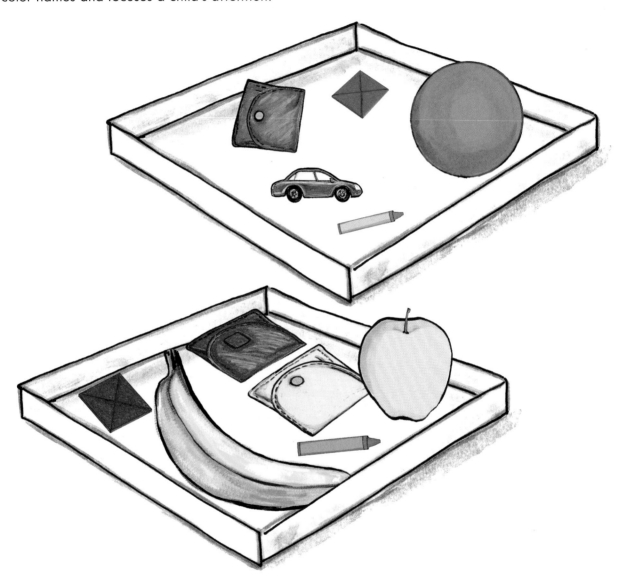

Look What We're Learning

Patterns & Algebra
- Sort objects by color

Sensory Motor
- Use visual cues to guide reaching for, grasping, and moving objects

Vocabulary

sort

Color Sort

Materials/Setup:

- 1 Basket filled with many small solid color objects. Can include: buttons, Tag Bags®, 4 Squares More Squares®, small blocks, counters, fruit, markers or crayons
- 2 Trays for sorting

Grouping:

Small group

English Language Learners:

As the objects are moved into groups, say a complete sentence to tell what is happening. **This ball is blue. Put it with the blue group. This chalk is not blue. Put it in the other group.** Have children use the same language to describe their own sorting.

Objective

Children sort objects by color.

Activity

Can you sort these objects into two groups?

1. **Take something from the basket. What is it? Is it blue?**

2. Each child takes one object from the basket.

3. **What is it? Is it blue?** Children put blue objects onto one tray. They put objects that are not blue on the other tray.

4. Return objects to the basket and sort by another color.

✓ Check for Understanding

Observe children as they sort the objects into two groups. Can they sort by color?

Support: Use Little Pieces from 4 Squares More Squares. These pieces are the same in every way except for color. Children can sort them into two groups: green and not green. Repeat with other colors.

More to Learn

Moving Colors

Children can also sort by color using their clothing. They can move into two groups. **If you are wearing blue, move onto the rug. If you are not wearing blue, move next to the windows.**

Counting Colors

Give each child a Little Piece from 4 Squares More Squares. Have them search for objects in the classroom that are the same color. Come together as a group and sort the objects they found by color.

Fit Test

Sort by Size

Children have begun to explore different ways to sort objects into groups by color. Now children will use size to sort objects into groups.

Look What We're Learning

Patterns & Algebra
- Sort objects by size

Problem Solving
- Use manipulatives to find a solution

Measurement & Time
- Compare size of two objects using big/small

Social-Emotional
- Take turns
- Work together to solve a problem

Vocabulary

big
little
size
small

Objective

Children use a shoebox with a small hole cut in top to sort toys by size.

Activity

Let's sort these items by size.

1. Show children a shoebox with a small hole cut in the top.

2. Ask children what objects they think will fit in the hole. Discuss big and little.

3. **Let's do the fit test!** Invite each child to select an object to see if it fits in the hole.

4. **You picked an apple. Does it fit? No, it doesn't fit. It must be too big.**

5. When children have tested, open the box and compare the little objects in the box with the big objects outside the box.

✓ Check for Understanding

Observe children as they sort toys by size. Can they sort by size?

Support: Exaggerate the size difference by giving children objects to sort that are the same in every way except for size, such as a small green blocks and big green blocks.

Materials/Setup:

- Shoebox with small hole cut in top
- 2 Trays or containers
- Assorted classroom objects or toys in varied sizes such as toy cars, blocks, coins, erasers, action figures, or stuffed animals

Grouping:

Pairs

English Language Learners:

Ask children to tell the words for big and small in their home language. This connects the activity to what they already know. Ask them to name some things that are big and some that are small. Have children repeat.

An elephant is big.
An ant is small.

More to Learn

Books in a Box

Repeat the activity to sort children's books by size. Will the books fit inside a shoebox with the lid on?

Funnel Fun

Repeat the activity using an upside-down funnel. Sort items that will fit through the small funnel hole.

Fastener Sort

Sort by Function

It is easier to sort by some attributes such as color. Other ways to sort are more subtle. When children sort Tag Bags® by fastener, they are able to focus on a less apparent attribute. Sorting by the type or kind of object encourages children to explore other ways to group or classify objects.

Look What We're Learning

Patterns & Algebra
- Sort objects by function/kind

Problem Solving
- Use manipulatives to find a solution

Social-Emotional
- Work together to solve a problem

Sensory Motor
- Use fingers to open and close fasteners

Materials/Setup:
• Tag Bags®

Grouping:
Whole class

English Language Learners:

Preview the names and types of fasteners before the activity. During the activity, when you call for each fastener type, hold up the bag.

Objective

Children sort objects by function.

Activity

How can we sort these Tag Bags?

1. Hand out one Tag Bag to each child. **We are going to sort the Tag Bags by the kind of fastener.**

2. **Try your fastener. Can you open and close your Tag Bag?**

3. **Find a friend who has your fastener. When you find a friend with the same fastener, sit down together.**

For more ideas, see "Fastener Sort" in the Tag Bags activity booklet.

✓ Check for Understanding

Observe children as they sort themselves into groups by fastener. Do they sort themselves into like groups?

Support: Begin with just three types of fasteners: loop, snap, and Velcro®.

More to Learn

Land, Sea, and Sky

Find objects that can be found in the sky, on land, or in water. Have children sort the objects by where they can usually be found (e.g., airplane in sky, fish in water).

How Do You Move?

Sort the 1-2-3 Touch & Flip® Animal Cards based on how each animal moves. Try out all the movements. Create piles for animals that swim (fish), fly (flamingo, quail, bird), walk (rhinoceros, turtle), crawl (spider, ants), and slither (snail, snake). Some animals may fit into more than one category!

Repeat After Me

Describe a Simple Pattern

Patterns are all around. Children can describe a pattern by describing the pattern unit. It is the part that repeats. Children learn the words used to describe patterns of color, shape, sound, and movement.

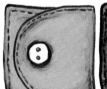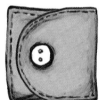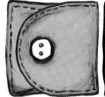

Look What We're Learning

Patterns & Algebra
- Identify and describe a pattern by telling the repeating unit
- Duplicate and extend pattern

Problem Solving
- Use manipulatives to find a solution

Social-Emotional
- Take turns
- Work with others to solve problems

Vocabulary

pattern
repeat

Repeat After Me

Materials/Setup:
• Purple and green Tag Bags®

Grouping:
Small group (4)

English Language Learners:
Children who are just learning English can benefit from watching other students add their pattern units first.

Objective
Children identify and describe a simple pattern.

Activity
Let's make a repeating pattern.

1. **Repeating is when you say or do something again. Repeat after me** use a robot voice: **Boop.** "Boop." **Bop.** "Bop." **Bleep.** "Bleep." **I want to make a pattern.** "I want to make a pattern."

2. **Good! Let's make a pattern. A pattern is something that repeats like this: green, purple, green, purple.**

3. Give each child one purple and one green Tag Bag. Place your green and purple Tag Bags on the table.

4. Have children repeat the pattern one by one, adding green then purple to the row. **When it's your turn say, I can repeat: green, purple.**

5. When all Tag Bags are on the table, read the whole pattern chorally, stressing the first unit of the pattern, **GREEN, purple, GREEN, purple.**

✓ Check for Understanding

Observe children as they create a simple pattern. Can they repeat the pattern? Can they extend it?

Support: Let children who do this activity easily take the first turns. It helps them to see others who go first.

More to Learn

Make a Pattern
Repeat the activity with classroom objects to make a pattern: crayon, block, crayon, block. Use objects that are the same color at first.

Bigger Pattern Units
For a challenge, make another type of pattern (ABB, AABB). Use the same pieces to have children build the pattern, or use instruments to make sound patterns.

Sounds Like This!

Duplicate a Simple Pattern

Children learn about patterns with many exposures and experiences. At first, use obvious patterns that are easy to repeat. Gradually, children will notice the part that is repeated. Then they can make their own with colors, shapes, sounds, movement, and numbers. Sound and movement patterns may come easily to some children, although they may be more challenging for others.

Look What We're Learning

Patterns & Algebra
• Duplicate and extend pattern

Social-Emotional
• Work together to solve a problem
• Listen to directions/cues

Vocabulary

pattern
repeat

Sounds Like This!

Materials/Setup:
- Wood Pieces:
  – 8 Big Lines
- *Sing, Sound & Count With Me* CD track 12, "Pattern Dance"

Grouping:
Whole class

English Language Learners:

As you say the names of the movements in the activity, have children chant along. **Tap, tap, tap; Clap, clap.** Children will see, hear, and move their bodies.

Objective

Children duplicate sound and movement patterns.

Activity

Let's find the pattern.

1. Sing and move along with the song "The Pattern Dance." Children can hear sound patterns and do the movements.

2. Seat the class in a circle and give every other child one Big Line. Those children tap three times. The other children use their hands to clap two times.

3. Direct children to repeat their sound and motion as you name the sound aloud:

   Tap, tap, tap
   Clap, clap

4. When children know their parts, they can repeat the pattern around the circle.

5. Have children switch their parts—Big Lines, clapping—and repeat the activity.

✓ Check for Understanding

Observe children as they tap and move. Can they duplicate the sounds and movements?

Support: Use a simple AB pattern with both a movement and a sound. Some children may be stronger auditory learners; others may be stronger kinesthetic learners.

More to Learn

Stringing a Pattern
Another variation is to make patterns by stringing beads. *I Know My Numbers* booklet 9 has a fun pattern activity using beads. Or, you can string your own with different shapes or colors to make a pattern.

Copy Cat
For a challenge, pairs can each make a pattern with blocks, shapes, or by coloring on paper. Then they swap and try to copy their partner's pattern.

Teacher, May I Grow a Pattern?

Explore Growing Patterns

There are different types of patterns. We have explored repeating patterns of color, shape, sound, and movement. Although repeating patterns simply repeat the unit (AB-AB-AB), growing patterns do more than repeat. They repeat and they add something to the repetition. They grow or get bigger with each repetition (A-AB-ABC-ABCD....).

Look What We're Learning

Patterns & Algebra
- Identify, describe, and extend a growing pattern

Problem Solving
- Use manipulatives to find a solution
- Look for a pattern to find a solution

Social-Emotional
- Take turns
- Use manners when speaking

Vocabulary

growing

pattern

repeat

Teacher, May I Grow a Pattern?

Materials/Setup:
- 4 Squares More Squares®

Grouping:
Small Group (4 children in a row facing teacher)

English Language Learners:
Children can respond with shorter sentences, such as "Teacher may I?" This encourages all children to participate. They may already be familiar with the word "grow"—with plants or children. Just like we grow bigger, the pattern gets bigger and bigger.

Objective
Children build growing patterns.

Activity
Let's make a growing pattern.

1. Face the children. **Growing patterns repeat and they get bigger every time they repeat. Look at the beginning of my pattern. How many squares do I have?** (One.)

2. Create another column. Repeat the previous column and add one square. **My pattern grew! How many do I have here?** (Two.) Repeat to create a third column.

3. Invite children to grow the pattern. Start with the child on your right. **Say to me, Teacher may I grow the pattern?** "Teacher may I grow the pattern?" **Yes, you may grow the pattern. Copy this one. Make it grow with one more Little Piece.** You can mix Big Pieces into the pattern when children are ready.

4. Repeat step 3 with every child.

For more ideas, see "What's My Pattern," "Checkerboard Quilts," and "Step by Step" in the 4 Squares More Squares activity booklet.

✓ Check for Understanding

Observe children as they grow the pattern. Can they identify the growing pattern? Can they describe the change?

Support: The children who need more support should have the first or second turn.

More to Learn

Double Dip
Give children cutouts of ice cream cones and scoops of ice cream. The first cone has one scoop, the second has two scoops. Have children glue three to five cones in a row on a large paper. Show a growing pattern that adds one each time. How many scoops will the fifth cone have?

Figure It Out
Show step 1 and step 3 of a growing pattern. Have the child create step 2.

Pattern Day

Explore Patterns in the Real World

Patterns are all around. Children learn the words used to describe patterns of color, shape, sound, and movement. They find the part that repeats and predict what will come next. They can see the repeating patterns on wallpaper, furniture, floor tiles, wrapping paper, and on their clothes.

Look What We're Learning

Patterns & Algebra
- Find and describe patterns in the world
- Identify and describe a pattern by saying the repeating unit

Problem Solving
- Look for a pattern to find a solution
- Draw a picture to help solve a problem

Sensory Motor
- Use fingers to hold crayons and other objects

Materials/Setup:

- On the day of this activity, have children come to school dressed in as many patterns as possible. They may find a striped shirt, plaid pants, or printed tights.
- Crayons
- Paper
- Pattern Day letter to families

Grouping:
Whole class

English Language Learners:

Review names of clothing (shirt, pants, socks). Children can talk about the patterns on their clothes in pairs. They can review their talk with a partner, then work together to tell the class about their clothes. Partners can be mixed, so English language learners do not need to speak in front of the group if they are not yet ready.

Objective
Children explore patterns in the real world.

Activity
Let's look for patterns on our clothes.

1. **Are you wearing a pattern today?**

   If NO, **I am not wearing a pattern today.**

   If YES, **I am wearing a pattern today. It has _____.** (stripes, design)

2. After children have described their patterns, have them draw a picture of the pattern on their clothes.

3. Invite children to share their drawings and read the pattern.

✓ Check for Understanding

Observe children as they look for patterns. Can they identify patterns in their environment?

Support: Have children work in pairs. Give each child five Little Pieces of the same color from 4 Squares More Squares®. Children take turns placing Little Pieces to make a repeating pattern.

More to Learn

Pattern Hunt
Look around the classroom. What patterns can you find? Some examples could include: floor tiles, windows, or blinds.

Nature Patterns
For a challenge, have children find materials in nature and make their own patterns. Children can describe the pattern unit (part that repeats).

MEASUREMENT & TIME
Short & Tall, Big & Small

Children begin to compare objects before they have the language to describe what they are doing. They can stack rings by size on a tower, and they can tell the difference between the baby's socks and their own socks. Determining the size of an object is called measurement. Pre-K children need to learn about size, width, length, weight, capacity, time, and the words we use to describe them (small, long, light).

Children learn basic concepts of measurement when they measure objects. They can measure by making direct comparisons between objects, using nonstandard units (cubes, paperclips), and using standard units (Burns, 2007). In Pre-K, we focus on direct comparisons and use of nonstandard units to teach children how and why we use standard measuring tools such as a ruler, scales, measuring cups, and clocks.

The activities in this domain allow children to build the following skills:
- **Make a Direct Comparison of Size**
- **Compare Length Using Long & Short**
- **Compare Height Using Tall & Short**
- **Compare Weight Using Heavy & Light**
- **Compare Width Using Narrow & Wide**
- **Compare Capacity Using More & Less**
- **Order By Size**
- **Explore Area**
- **Use Nonstandard Units of Measurement**
- **Sequence Events**
- **Connect Times & Events**

Below is the significant research for this domain. For additional Measurement & Time resources, see the reference section at the end of this teacher's guide.

Boggan, M., S. Harper, and A. Whitmire. 2010. "Using Manipulatives to Teach Elementary Mathematics." *Educational Research*. 3:1-6. http://www.aabri.com/manuscripts/10451.pdf.

Burns, M. 2007. *About Teaching Mathematics: A K-8 Resource*. Sausalito: Math Solutions.

Clements, D.H., Sarama, J., DiBiase, A. 2004. "Measurement in Pre-K to Grade 2 Mathematics." *Engaging Young Children in Mathematics: Standards for Early Childhood Mathematics Education*. Mahwah, NJ: Lawrence Erlbaum Associates, Inc.

Compare & Share

Make a Direct Comparison of Size

Young children compare objects to first understand measurement. They can look at two items and see that one is bigger, longer, or taller. They can hold two items and feel that one is lighter. They can pour water from one container to another to see which holds more. Long before they begin to measure using centimeters, pounds, or liters, they learn about measurement by using their senses to make comparisons.

Look What We're Learning

Measurement & Time
- Compare size of two objects using big/small

Problem Solving
- Use manipulatives to find a solution

Literacy
- Talk about observations
- Speak in complete sentences

Vocabulary

big

bigger

compare

small

smaller

Compare & Share

Materials/Setup:
- Big stuffed animal
- Small stuffed animal
- Wood Pieces
  – Big Lines

Grouping:
Small group (8)

English Language Learners:

Before children come back to share, say the name of their object. Have them repeat. Let English language learners listen to friends speak before sharing. It's okay to shorten the sentence: **This _____ is bigger.**

Objective

Children compare objects using big and small.

Activity

Today we are going to compare things. We'll find big things and small things.

1. Show big and small stuffed animals. **This bear is big. This bear is small. How do you know this bear is bigger?**

2. Listen to children's answers and encourage them to show with the stuffed animals. They may talk about what they can see or feel. They may talk about height, weight, or even how it feels to hug the bear (circumference).

3. Give each child a Big Line. Find something that is bigger than the Big Line.

4. **Let's compare the things we found. I'll go first. This book is bigger than my Big Line.** Hold the Big Line next to the book and show that it is bigger in both length and width.

5. Let each child share. Help them use a complete sentence: **This _____ is bigger than my Big Line.**

6. Repeat and find smaller things.

✓ Check for Understanding

Record children's answers about the size of the bear. Can they compare using big and small? You can use these as a baseline to show how their understanding of measurement builds throughout the year.

Support: Show children two objects, one smaller than the Big Line and one bigger. Compare the objects and the Big Line. Say if each object is bigger or smaller than the line. Then begin the lesson.

More to Learn

Family Comparisons
Have children bring photos of their families. Have them say who's bigger than they are and who's smaller.

Comparing Shapes
Vary the activity by passing out shapes and having children find a shape that is bigger/smaller from the center of the group.

Use Your Noodle

Compare Length Using Long & Short

Children know about length in a general way. They understand the difference between long and short hair, and long and short sticks. They need to learn how to compare objects and use longer and shorter to describe length. Line up the ends of each object to compare length. Children may not start that way. Appeal to their sense of fairness to explain why it is important that they start measuring from the same spot.

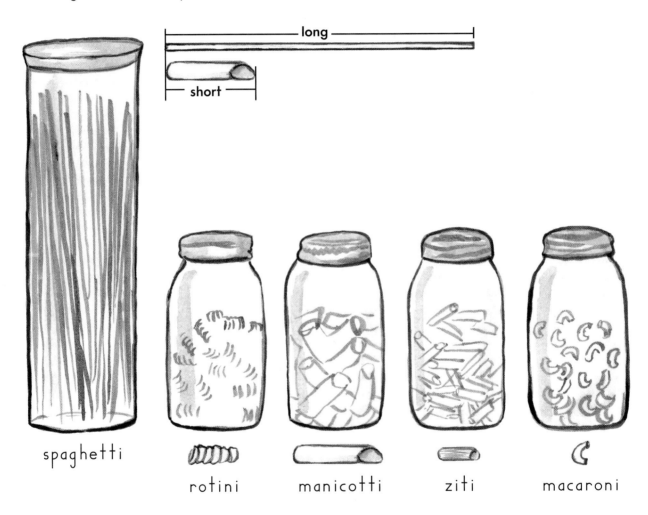

spaghetti   rotini   manicotti   ziti   macaroni

Look What We're Learning

Measurement & Time
- Compare length of two objects using long and short

Social-Emotional
- Work with others to solve problems

Literacy
- Repeat teacher's words

Vocabulary

long

longer

short

shorter

Use Your Noodle

Materials/Setup:

- 3+ Types of noodles (e.g., fettuccine, spaghetti, rigatoni, manicotti, ziti, farfalle)
- Small container for noodles

Grouping:

Small group, Whole class

English Language Learners:

Use Roll-A-Dough Letters® to demonstrate long and short. Show children a picture of a snake. **I will make a long snake.** Give children Roll-A-Dough. **Let's all make a long snake.** Have children repeat, "My snake is long." Repeat by making short snakes. Keep rolling the dough into longer and smaller sizes. Be sure to use hand gestures each time you say long and short.

Objective

Children compare length using **long** and **short**.

Activity

Today we'll play a game with long and short noodles.

1. Place a long noodle next to a short noodle. Line up the ends. **This noodle is long. I know it is long because it sticks out further than the other noodle. This noodle is short.** Point.

2. Let each child choose a noodle. Put children in pairs. **Let's wiggle like wet noodles. Wiggle with your partner.** Wiggle. **Stop!**

3. Have partners compare noodles. Help them line up the ends. **If your noodle is longer, raise it in the air! Repeat after me: My noodle is longer!** "My noodle is longer!" **Noodle, noodle, doodle, doodle.** "Noodle, noodle, doodle, doodle." **Noodles down.**

4. **If your noodle is shorter, raise it in the air! Repeat after me: My noodle is shorter!** "My noodle is shorter!" **Noodle, noodle, doodle, doodle.** "Noodle, noodle, doodle, doodle." **Noodles down.**

5. Help children find new partners. Repeat the activity.

✓ Check for Understanding

Observe as children compare their noodles. Do they compare noodles by placing them side by side and lining up the ends? Do they know which noodle is longer/shorter?

Support: If children have difficulty lining up the noodles to compare, give them a block to put at the end. Tell them to have both noodles touch the block.

More to Learn

Longer & Shorter?

Have children compare their noodle to their friends' noodles. Sometimes their noodle will be longer than their friends' and sometimes it will be shorter. Encourage them to think about and discuss why this could be.

Check the Line

For a challenge, have children make and compare lines with Tag Bags®. Give each partner a Tag Bag and have them look at the number inside. Have them build Tag Bag lines with that number of Tag Bags. Which is shorter? Longer?

Tall Towers

Compare Height Using Tall & Short

Children love to be measured. At home, families may record children's heights on a wall or doorframe. This activity builds on children's prior knowledge and gives them the language to compare by height.

Mathematically, the measurement of height is the same as the measurement of length. We want young children to know that height is measured vertically. Length is measured horizontally. Often, it is easier to accurately compare heights. As long as objects are on the same flat surface, they have a common starting point. Encourage children to measure from the bottom to get a fair measurement.

Look What We're Learning

Measurement & Time
- Compare height of two objects using tall and short

Problem Solving
- Guess and check the answer; repeat until correct answer is found

Social-Emotional
- Take turns

Vocabulary

height

short

tall

Tall Towers

Materials/Setup:

- 4 Squares More Squares®:
  - Little Pieces: 10 yellow, 10 orange, 10 purple, 3 red, 3 green, and 3 blue

Grouping:
Small group (6)

English Language Learners:
Review color words before you begin the activity. Have children point to each color and say the word.

Objective
Children compare height using tall and short.

Activity
Today we are going to build tall and short towers.

1. Demonstrate tall and short. Stand up. **Now I'm tall.** Crouch down. **Now I'm short.** Have children follow and repeat.

2. Mix up the Little Pieces in the center of the table. **We are going to build a tower for each color.**

3. Assign each child a color. Have them find their Little Pieces. **Which towers do you think will be tall?** Build towers.

4. **Who has a tall tower? Who has a short tower? Let's see.** Line up a tall tower side by side with a short tower. **Which one is taller? Which one is shorter? How do you know?** (The tall one comes up higher than the short one.) **Are any of our towers the same height?**

✓ Check for Understanding

Ask the children with tall towers to find a partner with a short tower. Ask children with short towers to find a partner with a tall tower. Can they find a partner?

Support: Demonstrate tall and short towers before children build. Point to the yellow tower. **The yellow tower is tall.** Point to the blue tower. **This tower is short.**

More to Learn

Taller Than Our Plant
Find a small potted plant. Carry the plant outside and find plants that are taller and shorter. Compare the plants directly by placing them side by side.

Taller Towers
Put children in pairs and have one select a 1-2-3 Touch & Flip® Counter Card. Have that child build a tower with the number of squares shown. Challenge the other child to make a taller tower. How many squares are needed?

Heavy or Light?

Compare Weight Using Heavy & Light

Children judge weight well because it is easy to feel. They can tell if there is juice in a cup by feeling its weight. Help them use math vocabulary to compare weight by picking up objects. To start, compare objects with very different weights, then move to objects that are closer in weight. This will make it easier for your children to understand weight when it is time to introduce scales and balances.

Look What We're Learning

Measurement & Time
- Compare weight of two objects using heavy and light

Social-Emotional
- Take turns

Vocabulary

heavy
light
weight

Heavy or Light?

Materials/Setup:

- Pairs of heavy and light items:
  - Empty milk jug and one filled with water
  - Choose your own. Make sure paired items are about the same size.

Grouping:
Small group (2-6)

English Language Learners:

Children can preview the objects before the activity. They can touch and play with them. Some common objects may be unfamiliar to children from different cultures. Name the objects and have children repeat. They may name them in English or in their home language.

Objective
Children compare objects using heavy and light.

Activity
We are going to hold things to learn about weight.

1. Put a pair of heavy and light items on each side of a large table. Group children in pairs.

2. Pass around the empty milk jug. **This jug is light.** Pass around the jug filled with water. **This jug is heavy.**

3. Send each pair of children to a different set of objects. **One partner should pick up one thing in each hand. Now switch. Tell each other which one is heavy. Now tell which one is light.** Listen to their answers and provide help as needed.

4. Have children move to a different set of objects and repeat.

✓ Check for Understanding

When children get to the last set, put out two hula hoops. Label one heavy and one light. Put an example in each hoop. Have each partner choose an object and put it in a hula hoop. Do they place the objects in the correct hoops?

Support: Adapt the activity to fit children's mobility needs. If it's not possible to pick up the object, place objects on a child's lap or another part of the body that can feel weight.

More to Learn

What's in Your Box?
Vary this activity by putting a different number of Tag Bags in two shoe boxes. Pass the boxes around. Discuss which is heavier and why.

What's in Your Bag?
Fill quart-sized bags with different objects. Try packing peanuts, marshmallows, grapes, or lug nuts. This will help children understand that objects of the same size can be different weights.

Let's Decide, Narrow or Wide?

Compare Width Using Narrow & Wide

This activity uses the block area to teach children about width. When they build a stall for a horse, a house for a dog, or a garage for a car, they're judging width. They make the space wide enough for the animal or vehicle to fit. In each of these examples, children also judge the element of height. Focus on width in this lesson by building a bridge (flat surface). Use children's constructions to reinforce measurement concepts and introduce vocabulary.

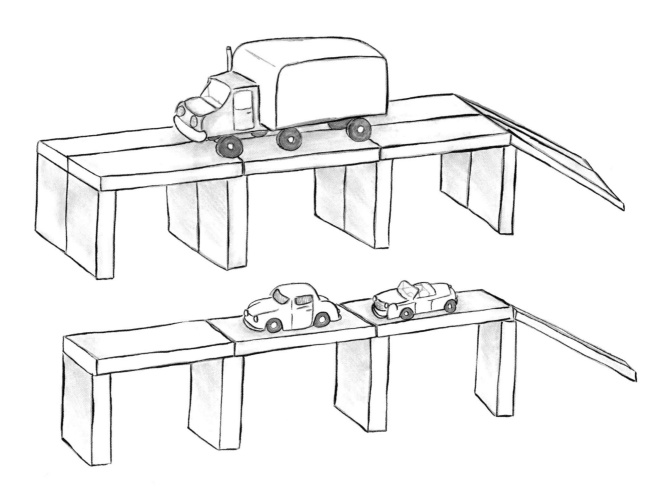

Look What We're Learning

Measurement & Time
- Compare width of two objects using narrow and wide

Problem Solving
- Guess and check the answer; repeat until correct answer is found

Sensory Motor
- Move object in hand to position for placement
- Use visual cues to guide reaching for, grasping, and moving objects

Vocabulary

narrow

wide

Let's Decide, Narrow or Wide?

Materials/Setup:
- Blocks
- Wide toy car/truck
- Narrow toy car/truck

Grouping:
Small group

English Language Learners:
Provide two sponge paintbrushes—one narrow and one wide. Invite children to make a painting with the two brushes. Point out narrow and wide lines. Have children show their paintings and say "narrow" and "wide."

Objective
Children compare spaces using narrow and wide.

Activity
Let's build a bridge that is wide enough for a car.

1. Give children the narrow car. Help them build a block bridge just big enough for the narrow car. **It fits! The bridge is wide enough.**

2. Pull out the big truck. **Can this truck fit on the bridge?** Listen to their guesses. Try it. **No! The bridge is too narrow for this big truck.**

3. Help children build a wider bridge. **Is this bridge wide enough for the big truck? It is! Can the little car fit?**

4. Encourage children to describe their constructions. Help them use wide and narrow.

✓ Check for Understanding

Playfully ask children to put a car on the wide bridge or the narrow bridge. Can they identify narrow and wide?

Support: Create narrow and wide spaces using classroom tables or chairs. Have children move through the spaces to feel narrow and wide.

More to Learn

Size Sort
Extend this activity by testing more vehicles on the original bridge. Sort into two groups: narrow and wide.

Word Time™
Use the Transportation Lesson from Word Time (pages 256–257) in our Language & Literacy program to explore wide and narrow.

What Holds More?

Compare Capacity Using More & Less

Children love to work with capacity. They may not realize it, but all the scooping and pouring they do at the water table or in the sandbox are experiences with capacity. We teach them to use holds more and holds less to compare the capacity of containers.

Look What We're Learning

Measurement & Time
- Compare capacity of two containers using holds more and holds less

Patterns & Algebra
- Sort objects by size

Social-Emotional
- Cooperate with other children
- Take turns

Sensory Motor
- Move an object in one hand to position it for use

Vocabulary

holds less

holds more

What Holds More?

Materials/Setup:

- 1 Medium-sized cup per pair

- 4 Containers of different sizes and shapes per pair

- Rice, water, or sand

- Sand/water table

Grouping:

Pairs

English Language Learners:

Choose items to measure that students are familiar with from their daily activities at home or school. If the containers have specific names (milk carton, cereal box), have children listen and say the names before the activity begins.

Objective

Children compare the capacity of containers.

Activity

Today we will fill some containers. Let's figure out which ones hold more and which ones hold less.

1. Give each pair a medium-sized cup (use same-sized cups for each group). **This is your cup. Use it to fill the other containers.**

2. Have one partner choose an empty container. The other should completely fill the comparing container with rice. **Dump the rice into the other container. Does the rice fit?** If yes, **That container holds more than your cup.** If no, **That container holds less than your cup.**

3. Have partners switch roles and pick another container. **Does this container hold more or hold less than your cup?**

4. Repeat for remaining containers.

5. Encourage children to put the containers in two piles—holds more and holds less.

✓ Check for Understanding

Observe children as they choose and fill containers. Can they tell which container holds more or holds less?

Support: Use containers that are very different in size to begin. It will be more obvious which holds more and which holds less.

More to Learn

Explore More

Set up a discovery station at a water table for children to pour liquids between various containers. Add interest by changing the material to be poured into containers. You can make the material festive (black/orange rice for Halloween, red confetti for Valentine's Day).

Test It!

For a challenge, compare a tall, narrow container with a short, wide container. Make sure the wide container holds more. Have children guess which holds more, then test.

From Bears to Chairs

Order By Size

It is important that children learn to recognize and arrange objects by size. Young children practice spatial reasoning by ordering objects from largest to smallest. This helps them with a range of activities from building more solid block towers to writing their names. Ordering by size also prepares children for more complex math such as putting numbers in order (1,2,3) and problem solving.

Look What We're Learning

Measurement & Time
- Order three objects by size

Literacy
- Listen to perform a task
- Retell a story with pictures

Sensory Motor
- Move an object in one hand to position it for use
- Use fingers to hold scissors

Social-Emotional
- Take turns

Vocabulary

big
little
medium

From Bears to Chairs

Objective

Children put objects in order by size.

Activity

Let's read a story about size.

1. Read *Goldilocks and the Three Bears.* Talk about the size of each bear.

2. Look at the size of the bears in *I Know My Numbers* booklet 3. **Who is the big bear? Who is the medium-sized bear? Who is the little bear?**

3. Look at the size of the chairs in *I Know My Numbers* booklet 3. **Who sits in the big chair? Who sits in the medium-sized chair? Who sits in the little chair?**

4. Use the cutouts from A Click Away and point to the beds. **Where does Papa Bear sleep? Where does Mama Bear sleep? Where does Baby Bear sleep?**

5. **Let's cut out the bears. Put the bears to bed. Paste Papa Bear in the big bed. Paste Mama Bear in the medium bed. Paste Baby Bear in the little bed.** Help as needed.

✓ Check for Understanding

Observe and listen to children's choices and comments when they talk about the story. Do they put the bears and beds in the correct order by size?

Support: Introduce only two characters first—Papa Bear and Baby Bear. Help the children notice who is big and who is small. Then, add Mama Bear and help children understand where she belongs in the order. Also, practice with the beds. Introduce only Papa and Baby Bear's bed before you add Mama Bear's.

More to Learn

From Bears to Squares

Have children practice ordering shapes by size. Give them three squares from Mix & Make Shapes™ and have them order by size. Try with other shapes. Have students order four or five objects to extend the activity.

From Bears to Buttons

For more challenge, give children three objects of different sizes and have them figure out which would go to each bear. You can use three buttons, paperclips, or books and have them place the object into a box for Papa, Mama, and Baby Bear by size.

Cover & See

Explore Area

When you think of area, you probably have scary formulas in your head (length x width, lw=A). Long before children learn formulas, they need to understand the concept of area. Young children get an exposure to area when they cover a shape with squares. The number of squares tells how big the area is. It will be easier for your children to learn area formulas in future years if they understand the concept.

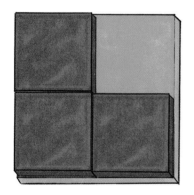  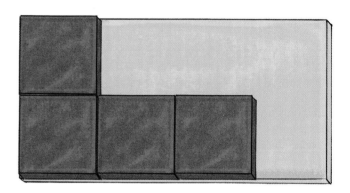

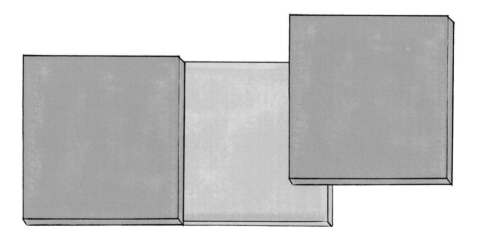

Look What We're Learning

Measurement & Time
- Cover an area with shapes to explore area
- Compare size of two objects using big and small

Number & Operations
- Verbally count a set of five objects

Problem Solving
- Guess and check the answer; repeat until correct answer is found

Vocabulary

cover

large

medium

not enough

small

too many

Materials/Setup:

- Mix & Make Shapes™:
  - 2 Large rectangles
  - 2 Medium squares
  - 8 Small squares

  (Add 4 Squares More Squares® Little Pieces if you need more.)

Grouping:

Pairs (4)

English Language Learners:

Before the activity, give each child a small, medium, and large square. Hold up a small square. **This is small.** Have children repeat aloud. Do the same with the other sizes. Then call out a size and children can hold up the matching square.

Objective

Children explore area by covering a shape with squares.

Activity

Let's find out how big our shapes are. We can cover them with small squares to find out.

1. Start with the medium square. **Let's cover this medium square. How many small squares do we need?** Let children take the number of squares they think are needed.

2. Invite one child to cover the medium square. Count the number of small squares. Children will want to compare the actual number with their guesses. Help them do this using **too many** or **not enough.**

3. **Now let's cover the large rectangle. How many medium squares do you think we need?** Let children take squares. Encourage them to explain their thinking.

4. Invite a child to cover the large rectangle. Count. Compare guesses. Repeat using medium squares. Have children try independently.

✓ Check for Understanding

Observe children as they cover the squares. Can children cover an area independently? Do any of the pieces overlap or stick out beyond the edges?

Support: Count out loud as each square is placed. Say, **one, two, three, four.** Ask, **How many squares did it take? It took four.**

More to Learn

Different Ways

Invite children to find out how many different ways they can cover one of the big shapes. Have them start using only the same shapes to cover. Then let them mix shapes to cover. Look at possible solutions on A Click Away.

How Many Squares?

Explore area using 4 Squares More Squares Big Pieces and Little Pieces. See "How Many Squares?" in the activity booklet for the full activity.

Measuring With Tag Bags®

Use Nonstandard Units of Measurement

When children understand the concept of measurement, they are ready to move beyond direct comparison. The next step is to use nonstandard units, such as paperclips, blocks, or Tag Bags. Children do this to become familiar with what it means to measure an object. They practice placing the units end-to-end, next to the item they are measuring. This activity will prepare them to transition easily to standard measures such as rulers and yardsticks in later years. You may want to add these standard measures to activity centers for children to explore.

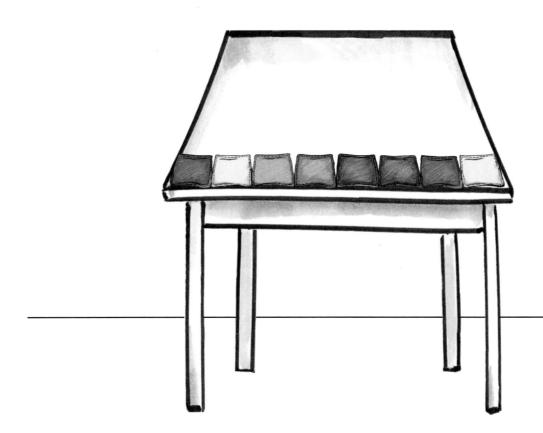

Look What We're Learning

Measurement & Time
- Use uniform objects (nonstandard units) to measure

Number & Operations
- Verbally count a set of 15 objects

Problem Solving
- Use manipulatives to find a solution

Vocabulary

long

row

Measuring With Tag Bags®

Materials/Setup:
- Classroom object (e.g., bookshelf, table, window)
- Tag Bags

Grouping:
Small group, whole class

English Language Learners:
Choose familiar items from home or school to measure. Children can benefit from whole or small group experiences. They can model their actions from their peers and increase participation.

Objective
Children measure an object using nonstandard units.

Activity
We are going to use Tag Bags to measure how long things are.

1. **Has anyone ever measured you? How did they measure you? What did they find out?**

2. Use Tag Bags. **Let's measure the table. Let's see how long it is with Tag Bags.**

3. Put the first Tag Bag at the edge of the table. **I'm going to put the first Tag Bag here at the edge. We can add Tag Bags to make a row on the table.**

4. Give each child a Tag Bag. Have them add their Tag Bags to the row, one by one. Make sure that the Tag Bags are touching at the edge.

5. Continue across the table. **When do we stop? We stop when we reach the edge. How can we find out how many Tag Bags are on the table? Let's count! This table is eight Tag Bags long.**

6. Repeat with other objects in the room.

See "Measuring" in the Tag Bags activity booklet for a fun song to sing during this lesson.

✓ Check for Understanding

Observe children as they measure with Tag Bags. Do children start at the beginning of the object? Do they place bags end to end?

Support: Demonstrate how to place Tag Bags in a row first. Then invite children to repeat and measure the object. They will benefit from seeing that the size of the object stays the same even when a different person measures.

More to Learn

More Squares
Use 4 Squares More Squares® Little Pieces to measure different objects in the classroom. **Can you find anything that is 5 Little Pieces long? 10?**

Measure Volume
Adapt this activity to measure the volume of a bottle or bucket. Have children use a smaller container to fill a larger one. Help them count the number of times they pour the smaller unit into the larger one.

Step By Step

Sequence Events

Young children have an understanding of sequence, which is the order of events. They know that they wake up before they eat breakfast, that they eat dinner before they go to bed, and that they take off their clothes before they take a bath. Children need to practice using words such as first, next, and last to describe the steps of simple tasks. By teaching sequencing, you prepare children to give directions, understand stories, build timelines, and understand how one event is related to another.

Look What We're Learning

Measurement & Time
- Sequence events in time

Problem Solving
- Act out problem to find a solution

Social-Emotional
- Demonstrate self-care skills

Vocabulary

first

last

next

Step By Step

Materials/Setup:
- Shoes
- Socks

Grouping:
Whole class

English Language Learners:
Have children say the word for first, next, and last in their home language. This will reinforce their understanding of the words in English.

Objective
Children sequence the steps of putting on socks and shoes.

Activity
I need to put on my shoes and socks. You can help me figure out what to do first.

1. Begin by tying your shoes, then try to put them on. **First, I'll tie my shoes. Next, I'll put my shoes on.** Accept children's corrections.

2. **I have to put the shoes on before I tie them? Okay.** Put the shoes on and tie them. Don't put the socks on yet.

3. Begin to put on your sock over your shoes. **Great! I have my shoes on. I'll put my socks on last.** Let children correct you.

4. Start from the beginning and ask the children to help you. **What should I do first? Next? Last?** Encourage the children to speak in complete sentences and use first, next, and last to describe the order.

✓ Check for Understanding

Listen as children sequence the steps. Can children tell you what happens first, next, and last?

Support: Demonstrate the correct steps first. Then try the silly version.

More to Learn

Mixed-Up Snack
At snack time, mix up the order of setting the table, eating, and cleaning up. Have the children correct you by telling what comes first, next, and last. Add a challenge by having more than three steps.

Line It Up
Read "Growing Pumpkins" or "How a Butterfly Grows" from the Line It Up™ story cards in our Language & Literacy program. Then have the children put the story in order. Teach first, next, and last.

Day & Night Charades

Connect Times & Events

Before children can tell time, they have a sense of the parts of a day and what happens in their lives at those times. They know that they play during the day and sleep at night. Children can describe the relationships between times and events with words such as morning, day, afternoon, and night. They can also use words, such as before and after to describe sequence.

Look What We're Learning

Measurement & Time
- Connect events with general times, such as day or night

Social-Emotional
- Participate in imaginary play
- Take turns with peers

Sensory Motor
- Naturally move and place body to perform tasks

Vocabulary

day

night

Day & Night Charades

Materials/Setup:
- Pictures from home
- Family Letter

Grouping:
Whole class

English Language Learners:
Preview two cards, one for day and one for night. Act out the picture on the card with children. Have one or two children take turns acting it out.

Objective
Children tell if something happens in the day or at night.

Activity
What do we do during the day? What do we do at night?

1. Place the day and night cards on the floor. Stack the pictures from home face down.

2. **Let's play a game. I am going to pick a picture. I will act out what is in the picture without talking. You guess what I am doing.**

3. Pick a picture and act it out. Give verbal clues if needed. When children guess correctly, ask the child who brought the picture if this activity happens during the day or at night. Place the picture under the correct time of day. Talk about some activities that happen during the day and night.

4. Pick students to act out their pictures. Have children whisper what is on the card to make sure they understand. Repeat.

✓ Check for Understanding

Observe children as they act and guess the time of day. Do they put the cards under the correct time of day?

Support: Review one or two picture cards in advance. Have children say if it's a daytime or nighttime activity. Make sure they get one of these cards during the activity.

More to Learn

Things We Do
Add picture cards for things that could be done in the day or night, like brushing teeth, reading books, and taking a bath. Make two copies of the cards and put in both day and night lists. You can also create a Venn diagram.

What Do You Do?
Have children create their own cards using photos or drawings. Play charades again and have children act out their own cards. Add their cards to the list.

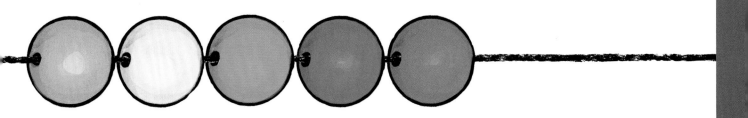

DATA REPRESENTATION & PROBABILITY
We Can Show What We Know

Data representation activities help children to collect and organize information (answers to questions) in a visual way. They answer questions about their everyday lives and explore how their individual answers become part of the whole-class response. Our activities offer children the opportunity to participate in survey questions in different ways. They begin by moving themselves to answer a question, then they learn to place an object or picture to show their response.

When children have collected and shown information on a graph, they are ready to think about what they have gathered. Repeated exposure to ideas and data in pictographs improves their understanding of comparison questions (Van De Walle, 2004). We ask children to think about the data they collected by asking them to compare the answers.

Probability is linked to data analysis. Probability can help us answer questions about our world with regard to how likely future events are. It helps children make sense of their day and their world. As they explore these ideas, they learn decision-making skills.

The activities in this domain allow children to build the following skills:

- **Move to Answer Questions**
- **Graph with Objects**
- **Explore Pictographs**
- **Identify Events as Likely or Unlikely**

Below is the significant research for this domain. For additional Data Representation & Probability resources, see the reference section at the end of this teacher's guide.

Newcombe, N.S., and A. Frick. 2010. "Early Education for Spatial Intelligence: Why, What, and How." *Mind, Brain, and Education* 4(3): 102-111.

Rudd, L. C., M. Lambert, M. Satterwhite, and A. Zaier. 2008. "Mathematical Language in Early Childhood Settings: What Really Counts?" *Early Childhood Education Journal* 36: 75-80.

Van De Walle, J. A. 2004. *Elementary and Middle School Mathematics Teaching Developmentally.* Boston: Pearson.

Clothes Lines

Move to Answer Questions

Children can answer a question with words or by moving into a group. Moving into groups helps children see answers. They can count how many or see the size of each group. Moving into groups is a way to organize information and answer questions.

Look What We're Learning

Data Representation & Probability
- Move to a designated location to indicate response to a question

Number & Operations
- Verbally count a set of 10 objects
- Compare sets of objects using more and fewer

Measurement & Time
- Compare length of two objects using long and short

Social-Emotional
- Work together to solve a problem
- Name body parts

Vocabulary

bottom
compare
longest
most
shortest
top

Clothes Lines

Materials/Setup:
- No materials needed

Grouping:
Whole class

English Language Learners:
Ask questions with inflection and hand gestures. Point to bottoms of clothing as you name them.

Objective

Children respond to questions by moving their bodies to different areas.

Activity

We're all wearing clothes. Let's look at all our different clothes.

1. **What type of clothes are you wearing on the bottom? Let's look. Are they pants? Skirts? Shorts?**

2. Have children move into groups by type of clothes. Show where each group should meet.

3. **Which group has the most children? Let's make lines to find out.** Children line up by type of clothing. **We can see which line is longest and which line is shortest.**

4. **We can also count the people in each line. Let's count to compare the number of children in each group. Which group has the most children?**

✓ Check for Understanding

Observe children as they move into groups. Do they choose the correct group? Do they know which group has the most?

Support: Begin with only two choices for bottoms (e.g., pants and no pants). Help children compare the two groups.

More to Learn

Movement Sort

Have children stand up if they have shoes with laces. They can sit on the rug if they have shoes with Velcro®. Then can move to the front of the room if they have buckle, slip-on, or any other type of shoe. Children can count to compare the groups.

Square Dance

Pass out 4 Squares More Squares® pieces. Allow children to move into groups by shape, type, and/or color. Have them compare the group size.

Apples & Bananas

Graph with Objects

When children have organized information using their bodies, they are ready to make simple graphs. At first, they will use concrete objects. They need to see, feel, and move the objects to understand how graphs work. Children will organize information visually in a graph to answer questions.

Look What We're Learning

Data Representation & Probability
- Represent data using concrete objects in a simple graph

Number & Operations
- Verbally count a set of 15 objects
- Compare sets of objects using more and fewer

Measurement & Time
- Compare length of two objects using long and short

Vocabulary

compare

fewer

longer

more

shorter

Materials/Setup:

- *Sing, Sound & Count With Me* CD track 7, "Apples & Bananas"
- Yellow bananas
- Red apples
- Poster paper with one column and two rows pre-drawn
- Letter to Parents

Grouping:
Whole class

English Language Learners:

Hold up the fruit and say the name. Let children pass around the fruit one at a time as you discuss it. Point out one similarity and one difference between the two fruits. Allow children to feel and smell the fruit.

Objective
Children place objects in a simple graph to answer questions.

Activity
What is the class favorite? Apples or bananas?

1. Show children an apple and banana. Talk about what they notice (shape, color) and what they like (smell, taste) about each. Have children choose their favorite.

2. Listen to "Apples and Bananas." Have children raise their fruit in the air when they hear its name.

3. **It's time to make a graph. A graph will show us which fruit our class likes best.** Children place their fruit in two horizontal rows. Compare the rows using: **more, fewer, longer, shorter.**

4. Have children find which fruit their class liked best. **Which row is longer? The longer row shows the favorite fruit.**

✓ Check for Understanding

Observe children as they place their fruit on the graph. Can they tell which fruit is the class favorite by comparing the rows?

Support: Use Little Pieces from 4 Squares More Squares®. Give children a pile of two colors of Little Pieces. Have children sort them by color and then line them up in rows. **Which row is longer?**

More to Learn

Compare by Pairs
For a variation, give pairs of children objects to sort. Have them line up the objects to compare.

More to Eat
For a challenge, give children a choice of more types of fruit or other snacks. Have them make a graph with three or five rows.

Dog or Fish?

Explore Pictographs

Pictographs are graphs that use pictures. Lining up pictures in a graph organizes information.
After children have made graphs with objects, they can make graphs with pictures. Using pictographs
prepares children for graphs they will encounter later in school—bar graphs and line graphs.

Look What We're Learning

Data Representation & Probability
- Represent data using pictures in a simple graph

Number & Operations
- Verbally count a set of 15 objects
- Compare sets of objects using more and fewer

Measurement & Time
- Compare length of two objects using long and short

Social-Emotional
- Work together to solve a problem
- Take turns

Vocabulary

fewer

longer

more

row

shorter

Materials/Setup:

- Two-row horizontal graph labeled with a dog and a fish

- Dog and fish stickers or cutouts

A Click Away
getsetforschool.com/click

Grouping:
Whole class

English Language Learners:

Have children make rows with blocks. Model using comparison words (longer, shorter) to describe the rows. Have children repeat your sentence or say one of their own using comparison words.

Objective
Children make and analyze a pictograph.

Activity
Which pet is the class favorite? Dog or fish?

1. **What is your favorite pet? A dog or a fish?** Children take a sticker to show their choice.

2. **What is the class favorite? A dog or a fish? Let's make a graph to see the class favorite.** Children take turns placing their stickers on graph. **A sticker can show my choice. Stickers in rows can show everyone's choices.**

3. Have children compare the length of the rows to find the class favorite. **Rows are organized. Which row is longer? Which row has more? The rows help us see the favorite pet.**

✓ Check for Understanding

Check that children place their sticker in the correct row. Can they identify the class favorite? Do they know which row is longer and which row is shorter?

Support: Use 1-2-3 Touch & Flip® Animal Cards to practice counting and comparing. Have children choose two cards. Count the animals on each card. **Which has more?**

More to Learn

Class Vote
You can use this activity any time you have a class vote. Place each child's picture in a magnetic photo frame. When the class has a decision to make, have children place their pictures next to their choice.

Choice Challenge
Add complexity by giving children more than two choices. Or, change the orientation of the graph, showing data in columns, from the bottom up.

Snow Boots or Sunglasses?

Identify Events as Likely or Unlikely

Children can make good predictions. They can use past experience to decide if an event is likely or unlikely. They will explore the likelihood of outcomes in both games and real life.

Look What We're Learning

Data Representation & Probability
- Discuss real world events as either likely or unlikely

Social-Emotional
- Demonstrate self care skills: dressing
- Take turns

Vocabulary

certain
impossible
likely
unlikely

Snow Boots or Sunglasses?

Materials/Setup:

- Large cutouts of a sun, clouds/rain, and snow

- Pictures or actual clothing items for different weather (bathing suit, mittens, rain hat, boots, scarf, and sandals)

Grouping:
Whole class

English Language Learners:

Have children preview the clothes they will use in the activity. Say, **When something is likely, it is like nodding your head up and down, which means Yes. When something is unlikely, it is like shaking your head right and left, which means No.**

Objective
Children predict the likelihood of an outcome.

Activity
Which clothes would you wear in different weather? Let's use the words likely or unlikely to talk about what you would wear.

1. **It is hot. You are likely to wear. . .** Show sandals. **You are unlikely to wear. . .** Show boots.

2. **It is raining. Are you likely or unlikely to wear/use _____?** Show item.

3. **It is snowing. Are you likely or unlikely to wear/use _____?** Show item.

4. Have children take items and go around the circle and say when they would wear or use them. They can choose a type of weather and put on the clothes and accessories that would likely be worn. They can take turns saying what they are wearing and why.

✓ Check for Understanding

Observe children's choice. Can they explain why they made their choice using likely or unlikely?

Support: Talk about likely and unlikely as it is really happening in the classroom. **We are putting on our coats at the end of the day. Is it likely that we will go home next?** Repeat often to reinforce this skill.

More to Learn

Spinning to Predict
Another variation is to use games. Make spinners using a paper plate and a fastener to hold a paper arrow. Making one section much bigger than the other, have children predict where it will land. The areas can be colors, shapes, or pictures.

Certain or Impossible
For a challenge, ask other reasoning questions about real events. Add the words **certain** and **impossible** to the discussion. **It is certain I will wash my hands before eating my snack.**

RESOURCES

Numbers & Math Benchmarks

Number & Operations

- Match one-to-one up to 5, 10, 15 objects
- Verbally count a set of 5, 10, 15 objects
- Recognize that the last number said is the total
- Recognize that totals are not affected by order or arrangement
- Make or draw a set of objects to match a given number
- Say at a glance how many are in a set up to three without counting
- Say an object's position in a line using ordinal numbers
- Compare sets of objects using more and fewer
- Identify written numerals and position correctly
- Connect numerals to quantities they represent
- Write numerals up to 10
- Count and label sets up to 10
- Combine sets to find out how many in all by counting up to 10
- Take objects away from a set to find out how many are left by counting
- Share a set of objects evenly with two or three classmates
- Divide one whole object into two equal pieces

Geometry

- Identify position or location using in and out
- Identify position or location using before and after
- Identify position or location using top, middle, and bottom
- Identify position or location using above and below, over and under
- Identify position or location using left and right
- Match shapes of same size, shape, and orientation
- Recognize that shapes can be the same even if sizes differ
- Recognize that shapes can be the same even if positioned differently
- Identify and describe circles
- Identify and describe rectangles
- Identify and describe squares
- Identify and describe triangles
- Identify specific shapes within a group
- Identify shapes in real world objects
- Recognize three-dimensional shapes

Patterns & Algebra

- Identify objects or pictures as the same or different
- Sort objects by color
- Sort objects by size
- Sort objects by shape
- Sort objects by function/type
- Identify and describe a pattern by saying the repeating unit
- Duplicate and extend pattern
- Identify, describe, and extend a growing pattern
- Find and describe patterns in the world

Measurement & Time

- Make a direct comparison of size
- Compare size of two objects using big/small
- Compare length of two objects using long/short
- Compare height of two objects using tall/short
- Compare weight of two objects using heavy/light
- Compare width of two objects using narrow/wide
- Compare capacity of two containers using holds more/holds less
- Order three objects by size
- Cover an area with shapes to explore area
- Use uniform objects (nonstandard units) to measure
- Sequence events in time
- Connect events with general times, such as day or night

Data Representation & Probability

- Move to a designated location to indicate response to a question
- Represent data using concrete objects in a simple graph
- Represent data using pictures in a simple graph
- Answer questions using organized data
- Discuss real world events as either likely or unlikely

Problem Solving

- Act out a problem to find a solution
- Use manipulatives to find a solution
- Guess and check the answer; repeat until correct answer is found
- Look for a pattern to find a solution
- Draw a picture to help solve a problem

Social-Emotional Benchmarks

Children in Pre-K have a wide variety of skills. Some may have exceptional motor skills, while others may excel in socialization. We can't assume that children will do well in all areas, which is why many of our activities (although specific to fostering early math, literacy, and writing skills) have social and motor components.

Social-Emotional

To develop social-emotional wellness in children, we must nurture:
- Innovation
- Responsibility
- Team work
- Perseverance
- Independence

Today, we have many young English language learners who need additional reassurance that they are safe and accepted. There may be things you notice about the ways these children socially engage with their peers depending on their culture. It's important for Pre-K teachers to understand that English language learners may misinterpret gestures and social interactions. For example, young Hispanic children may look down when speaking to an adult. In their culture, this often is a sign of respect. Regardless of their understanding of and ability to speak English, all four-year-olds need guidance and support to build strong social and emotional skills.

What are we teaching?

Our activities help children to develop self-concept, self-regulation, personal initiative, emotional understanding, and relationships with adults and peers. We want them to have positive self esteem, learn to engage in classroom activities, transition appropriately, take initiative, understand feelings, and take turns sharing and playing with their friends.

How do we do it?

- Model ways to show respect for self, people, things.
- Recognize, name, and respond to feelings. I can tell you are _____ because you are _____.
- Set class rules and teach children to follow them.
- Read stories that introduce children to values. The "Three Little Pigs" teach hard work and hard bricks save the day (visit the Book Connection section of this guide).
- Teach children how to help each other—to work together to lift and carry, to sweep into a dustpan, to say STOP when someone is hurting, to sit beside a sad friend.
- Participate in song, take turns, and share.

Social-Emotional Benchmarks

Self-Concept
- Demonstrate positive self esteem
- Demonstrate self-care skills like using the bathroom and dressing (putting on coats, washing hands) as age-/ability-appropriate
- Name body parts
- Describe changes in own body
- Imitate teacher's body movements

Self-Regulation
- Separate from caregiver without stress
- Manage emotions through negotiation and cooperation
- Respond to difficulty without harming self or others
- Manage and handle transitions well and without incident
- Understand and follow classroom routines
- Participate in clean-up routines working with other children
- Treat property with respect

Personal Initiative
- Demonstrate a desire for independence
- Show interest in many different activities

Emotional Understanding
- Name feelings he or she is experiencing
- Name emotions displayed by others
- Show empathy by offering comfort and help when appropriate

Relationships with Adults
- Interact easily with familiar adults
- Ask for help when needed
- Participate in conflict resolution activities (e.g., puppets acting out scenarios)

Relationships with Peers
- Cooperate with other children
- Participate in imaginary and dramatic play
- Take turns with peers
- Work with others to solve problems

Sensory Motor Benchmarks

Children will naturally engage in discovery and exploration. All you need to do is make materials accessible for them to play with freely. To develop sensory motor skills engage children in activities that promote:

- Movement
- Building and sorting
- Manipulation
- Processing sensations
- Organization

Our activities are hands-on so that children will experience sensory motor learning seamlessly in all that they do. As with social-emotional skills, some children will be better at sensory motor skills than others. Math is very hands-on with puzzles, blocks, nesting toys, pouring, and building. Our manipulatives have unique features to encourage motor development. For example, our Tag Bags® have six different fasteners. These fasteners naturally encourage a child's use of bilateral motor skills, and children love to open/close and reveal what's inside.

What are we teaching?

Our math activities encourage handedness, fine and gross motor movements, correct crayon grip, tool use, manipulation, motor coordination, motor planning, and body awareness.

How do we do it?

- Use music that encourages movement, finger plays, counting with fingers, clapping, and tapping.
- Set out manipulatives or selected pieces (Tag Bags, 4 Squares More Squares®, 1-2-3 Touch & Flip® Cards) and allow children to freely discover and explore.
- Allow children to hold books, turn pages, and point to letters and pictures in them.
- Use counters, blocks, and other items that can be counted, stacked, and sorted.
- Model proper tool use and manipulation—children may not always understand ways to manipulate pieces. For example, you may at times need to model how to flip a piece.
- Promote coloring as well as tracing letters and numbers.

Sensory Motor Benchmarks

Fine Motor

- Consistently use same hand for holding crayons, toothbrush, spoon, etc., and for performing skilled tasks
- Use fingers to open and close fasteners, hold crayons, scissors, cards, beads, etc.
- Move an object in one hand to position it for use, placement, or release
- Use index finger to trace letters or numbers on cards or in the air. Move fingers to show age/number and for finger plays

Gross Motor

- Use large muscle groups to maintain posture/position and mobility (i.e., walk, run, hop, skip, jump, climb stairs)

Bilateral Motor Coordination

- Use helping hand to stabilize objects and papers
- Use both sides of the body in activities (i.e., for drumsticks, other instruments, dancing)

Visual Motor Control

- Look at hands and use visual cues to guide reaching for, grasping, and moving objects

Body Awareness

- Know where the body is in relation to space
- Use enough but not too much pressure when holding and using tools
- Reach across midline to get an object from other side

Movement Perception

- Tolerate motion in activities; play with body awareness, balance, and regard for people and equipment

Touch Perception

- Handle play and art materials without an avoidance response
- Perceive the size, shape, or identity of an object by sense of touch

Visual Perception

- Notice and attach meaning to visual information

Motor Planning

- Move naturally and place body to perform tasks

School to Home Connection

Research consistently shows that a strong school to home connection helps children build self esteem, curiosity, and motivation to learn new things. For young children, home and school are the two most important places. A successful teacher-parent partnership assures children that there is a team of people who has their best interests at heart. Both parents and teachers provide a unique perspective about the child, so open, regular, and meaningful communication is important.

When teachers and parents team up, everyone wins!

Here are 10 ways to make a strong school to home connection:

1. Take advantage of opportunities to communicate during planned Pre-K events like teacher-parent meetings, conferences, and school visits. Take extra steps to communicate through letters, email, and even podcasts.

2. Share important assessment information about a child. Most schools have regular parent reports to share key progress details. Use our Get Set for School® Pre-K readiness assessments to help you identify what your children know and can do, and easily share information with families and other educators. You can find them at **getsetforschool.com**

3. Share your curriculum with parents. Show parents this Get Set for School Numbers & Math program. Let them play with some of the products. Tell them about our website, getsetforschool.com, so they can explore the many resources there. Encourage parents to ask questions about what their children are learning.

4. Share music with parents. Send children home singing songs from our *Get Set for School Sing Along* CD or our new *Sing, Sound & Count With Me* CD. Parents can share any fun song that they sing at home or ask to come in and teach it to the children.

5. Reinforce learning at home. Encourage parents to do finger plays and read books and nursery rhymes at home. Let them know just how important it is to their child's growth and development. Consider a teacher/parent sharing day where parents can learn some of the finger plays and reading strategies for children. For parents whose children's first language is not English, encourage them to sing songs and rhymes in their home language too.

6. Model language and thinking skills out loud. Children benefit from hearing adults talk and solve problems. They learn vocabulary and critical thinking skills. Parents, share your thoughts and ideas throughout the day. "It looks like it's going to rain outside. I'd better take an umbrella."

7. Read, read, read! Reading to children is fun and helps to build comprehension and language skills. Parents should read to their children regularly. It's good to read books repeatedly because this helps build memory and deepen comprehension.

8. Share *I Know My Numbers* with parents. Send each booklet home when children have completed it. Encourage parents to read it with their child and share it with other family members. Point out the page in each booklet that describes all the activities they can do at home.

9. Write, write, write! Help parents prepare their children to write. Encourage them to learn proper grip and support their child in holding a crayon correctly. Children love to see their names in print. Help them write their names in block capitals, which are the first and easiest letters to write.

10. Help children recognize letters and notice that print is all around them. Point out signs, logos, and letters wherever you go.

Dear Family Member,

This year we are using the Get Set for School® Numbers & Math program in our classroom. It's an exciting hands-on curriculum developed by a team of educators who understand young children and playful learning. The program is designed to teach young children the foundation for more complex math skills. The activities are so fun that the children just think they are playing!

The songs you may hear your child singing come from the *Sing, Sound & Count With Me* CD, which was recorded by Grammy-winning artists. Other unique Get Set for School products— flip cards, colored bags, and different shaped pieces, all made of high quality materials with dozens of unique features—are used to teach specific number and math skills.

Materials will be sent home throughout the year for you to review with your child. You may also be asked to send in items from home to be used in the classroom. To learn more about the Get Set for School curriculum, go to getsetforschool.com. There are parent resources, games, and downloads for practice and home use.

Sincerely,

Date _____

Dear Family Member,

In our Get Set for School® Numbers & Math program, we are currently learning about

_____.

We are using _____ to help us learn.

Please help your child choose a/an _____ to bring to school on

_____ for this activity.

Sincerely,

Books – An Inspiring Adventure

Books are an important part of a child's life and should be introduced as early as possible. Provide babies with chewable books, sturdy plastic, cloth, or board books for a great beginning. Books open a child's life to a world of adventure. Reading nursery rhymes and fairy tales to young children is an excellent opportunity for them to begin their own journey. As they read stories, children get to explore new places, discover new ideas, and learn about new and exciting characters. Early exposure to these stories sparks their imagination, increases their vocabulary, and ignites a lifelong joy of reading and learning. Books encourage children to be inquisitive, forward thinkers, and provide an amazing gift of knowledge.

Reading to young children inspires them to learn to read and continue reading more as they get older. Books should be read with enthusiasm and plenty of expression. Teachers and parents can help to foster the love of reading in children. Children love to be read to and really enjoy interactive reading which allows them to be a part. Being encouraged to point out items on a page, predict what will happen, repeat a rhyme, or even act out part of a story engages a child to read a book.

Fostering the Love of Reading

Parents can promote their child's love of reading simply by reading to their children at home. It can also come through building a home library, giving books as a gift/reward, and obtaining a membership card to a local library. A more creative way of fostering reading is to help children create their own books or tell/act out their own stories. When it's story time, let them pick a book. Parents should also ensure that children see them reading.

You should also read to children continuously. You can serve as a model for parents who struggle with reading. Children will get into books with a teacher who makes characters come alive and makes reading books a grand adventure. The use of entertaining voices, sound effects, gestures, and expression can make children love storytime and books.

Selecting Books

Selecting books for children can be a difficult task. First, it is good to understand the types of books that are available for a specific age range. Next, it is important to discover the types of books a child likes. Allowing children to browse books at the library and bookstore can help with selection. Websites, teachers, librarians, and other parents are also good sources for book recommendations. After choosing a book, browse and assess the book to determine if it's appropriate for your audience. Choosing the right book is an excellent start to a child's lifelong reading adventure. We have provided several options in the Book Connection section beginning on page 172.

Organizing Books

Books organized by subject are easier for children to quickly find the books that interest them. Try placing like subjects in bins around the room, and label them with a picture and print (e.g., place a picture of a Triceratops on the bin with the word dinosaur). Also, consider placing props in the bins. This will entice children to reenact the story using concrete objects. Encourage children to explore books and share their favorite books with their friends.

Encouraging Parents to Read

As you know, parents are their child's best teacher and yet, many parents do not read regularly to their children. Your advice and encouragement can make all the difference for parents. Remind parents regularly that reading books with their child every day will almost ensure that their child will love learning. Mem Fox, a literacy expert, has influenced and inspired our thinking about parents reading to children. Ms. Fox has published more than 30 books for children and taught literacy for more than 25 years. She believes that there is an important bond that develops as parent and child share in the joys and adventures of books. Ms. Fox also emphasizes that reading aloud to children every day is key to their later success in school and life.

Beyond helping the reluctant parent readers, you will also need to work with parents who don't know how to read. In this situation, one successful approach is to steer them to the librarians. Librarians are great resources for putting books and children together. At story times, librarians read to children, modeling for parents how to read to children of different ages and stages.

Book Connection

We recommend that you read related books to further expand the concepts in the Pre-K Numbers & Math program. Each book on the list entices children with simple, inviting text and clear, attractive illustrations. The books are an appropriate length for Pre-K read-alouds. We hope you enjoy reading them as much as we did.

Number & Operations

Henry the Fourth by Stuart J. Murphy. Illustrated by Scott Nash.
A dog show illustrates the concept of ordinal numbers. Four dogs take their turns performing first, second, third, and fourth until the winner is crowned. Suggestions for reinforcing math skills at home are included.

Mouse Count by Ellen Stoll Walsh.
A hungry snake catches 10 mice and counts them 1 by 1. They trick him into looking for another mouse and escape, counting down until none are left.
Also by Ellen Stoll Walsh: *Mouse Shapes.*

One Duck Stuck by Phyllis Root. Illustrated by Jane Chapman.
Groups of animals in a marsh come to help the duck who's stuck. They entertain children with their noises and encourage readers to count the animals in each group.

One Is a Snail, Ten Is a Crab by April Pulley Sayre and Jeff Sayre. Illustrated by Randy Cecil.
Animal feet engage children in counting. Children learn about animals with different numbers of feet. They see that the next number is always one more than the number before, a basic premise of addition.

One Moose, Twenty Mice by Clare Beaton.
Count from 1 to 20 with these friendly handcrafted animal pictures. Children will love searching for the cat that hides until the end of the book.

Geometry

Circle Dogs by Kevin Henkes. Illustrated by Dan Yaccarino.
Two long dogs curl up into circles when they sleep. When they wake up, the dogs show us other shapes around them and more of their favorites, circles.

Jack the Builder by Stuart J. Murphy. Illustrated by Michael Rex.
Children can identify with Jack as he builds his block creations. They count, recognize shapes, imagine amazing towers, and CRASH! Start all over again.

Mat Man Shapes by Jan Olsen. Illustrated by Molly Delaney.
As Mat Man's body changes into different shapes, children see the new things he can do. They learn the characteristics of shapes through joyful Mat Man adventures.

Round Is a Mooncake by Roseanne Thong. Illustrated by Grace Lin.
A little girl finds shapes in many objects around her home and in cultural celebrations.

Shapes, Shapes, Shapes by Tana Hoban.
Go on a shape hunt through each scene in this wordless book. Vivid photographs show us that shapes really are everywhere.
Also by Tana Hoban: *Cubes, Cone, Cylinders, & Spheres; Is It Larger? Is It Smaller?*

Patterns & Algebra

Lots and Lots of Zebra Stripes: Patterns in Nature by Stephen R. Swinburne.
Beautiful photographs take children on a pattern safari. Simple text explains patterns and
helps children see diverse patterns in the real world.

A Pair of Socks by Stuart J. Murphy. Illustrated by Lois Ehlert.
An unlikely main character, a single sock, helps children learn about patterns, sorting, and pairs as
it looks for its mate.
Also by Stuart J. Murphy: *The Best Bug Parade.*

Pattern Bugs by Trudy Harris. Illustrated by Anne Canevari Green.
The Pattern Bugs display their patterns in movements, sounds, and colors. Can you find all the patterns
hidden in this book? An instructional section gives a concise explanation of patterns.
Also by Trudy Harris: *Pattern Fish.*

Shoes Shoes Shoes by Ann Morris.
Matching pairs and sorting are underlying concepts in this book that takes us around the world.
Shoes introduce categories as well as faraway places and cultures.

Three Little Firefighters by Stuart J. Murphy. Illustrated by Bernice Lum.
How can the firefighters get ready without finding matching buttons to fix their coats? Children sort
by shape and size to help them out.

Measurement & Time

Biggest, Strongest, Fastest by Steve Jenkins.
Striking collages show us superlatives in the animal kingdom. Notes about each animal provide
the facts behind each statement.
Also by Steve Jenkins: *Actual Size*

Goldilocks and the Three Bears retold and illustrated by Jan Brett.
Comparisons have entertained children for countless years through this classic tale.

Inch by Inch by Leo Lionni.
A clever inchworm escapes a hungry robin by proving his valuable skill, measuring. He measures
and measures until he is safe.

Just a Little Bit by Ann Tompert. Illustrated by Lynn Munsinger.
Expressive animal characters rally around their elephant friend to help him play on the seesaw.
Everyone cheers when an unexpected friend gets the elephant off the ground.

Size by Henry Pluckrose.
Part of the Math Counts series, *Size* poses questions that encourage children to make comparisons.
Also by Henry Pluckrose: *Time, Length.*

Data Representation & Probability

If You Give a Mouse a Cookie by Laura Numeroff. Illustrated by Felicia Bond.
Children love the adventures of the mouse that wants to do one thing after another. They have fun predicting what the mouse might do next.

Let's Eat Lunch by Susan Vaughan.
An everyday event, lunch, teaches children about picture graphs.

Tally O'Malley by Stuart J. Murphy. Illustrated by Cynthia Jabar.
On their long drive, the O'Malleys pass the time by counting things they see. Tally's marks help them keep track of the score.

Two Dogs Swimming by Lynn Reiser.
These playful dogs encourage children to consider who is likely to win in each challenge. The story also subtly teaches a lesson about overcoming fears.

That's Good! That's Bad! by Margery Cuyler. Illustrated by David Catrow.
Each scenario in this book could be good or bad. Children will enjoy each fantastic situation. Each adventure can serve as a basis for survey questions. Children can indicate whether they think it is good or bad.

Social-Emotional

Cookies: Bite-Size Life Lessons by Amy Krouse Rosenthal. Illustrated by Jane Dyer.
Cookies teach children about important social concepts and vocabulary.

Elephant on My Roof by Erin Harris.
There's an elephant on Lani's roof, but at first, no one will help. After the neighbors finally help Lani, the elephant actually helps them in ways only an elephant can.

From Head to Toe by Eric Carle.
Animals demonstrate movements and help children learn body awareness. Children will want to move along with the animals in the book. But be careful with the donkey kicks!

It's Mine! By Leo Lionni.
Three frogs quarrel about everything until a storm floods their home. When they are saved, they realize that it's better to share.

Llama Llama Misses Mama by Anna Dewdney.
It's the first day of school and Llama Llama feels shy and sad. He misses Mama. His patient teacher and new friends help him see that he can love Mama and school.

Nursery Rhymes & Finger Plays

Favorite Nursery Rhymes from Mother Goose. Illustrated by Scott Gustafson.
Beautiful illustrations capture the children's imagination as they read these classic rhymes.

Hand Rhymes by Marc Brown.
Rhymes have motions indicated in small pictures alongside the text. Marc Brown's friendly illustrations engage both adults and children.

Head, Shoulders, Knees, and Toes and Other Action Rhymes by Zita Newcome.
This collection includes many old favorites and some less familiar rhymes. Small pictures along the sides of the pages show you the motions.

Mother Goose: Numbers on the Loose by Leo and Diane Dillon.
What a great idea to gather all the Mother Goose rhymes using numbers into one collection! Whimsical illustrations show the objects clearly, so children can count as they read.

Playtime Rhymes for Little People by Clare Beaton.
Intricate appliqué work illustrates this book of action rhymes. Each rhyme includes directions for motions to do along with your children.

Assessment

Families always want to know how their child is doing. Assessment helps you track what students know and can do. It allows you to share information about a child's learning with families, specialists, educators, and administrators.

Assessment can help you make good decisions for your children. Each assessment is a snapshot in time. Looking at those snapshots over weeks and months shows the path of a child's development and learning. This is the foundation for planning a curriculum—the set of experiences, activities, and materials you carefully select to help children learn.

We also assess young children to determine who would benefit from special services. Early identification helps families and schools address learning needs quickly.

Use Check for Understanding

Each activity in this guide focuses on a learning benchmark or skill. The Check for Understanding section helps you determine how well children understand or use the skill. We don't expect that your children will master each skill in one exposure. Young children need multiple opportunities to practice something before they really understand it. Use Check for Understanding to find out what your children understand so you can plan future experiences.

How We Check

In *Basics of Assessment: A Primer for Early Childhood Educators* (by McAfee, Leong, and Bodrova), the authors describe several ways to gather information about children. We use three of their methods in Check for Understanding.

Observe: Watch children as they move through an activity. What do they say? How do they move? What do they try? How do they solve problems? Collect this information so you can change course during the activity and make good decisions about future lessons.

> Rayshawn sorts his Tag Bags® by color. He sees that there are numbers inside and puts them in a line: 1, 3, 5, 4, 2. He counts to five, touching each bag as he counts. His teacher makes a note of this and plans a lesson on reading numerals.

Review Work Products: Children create drawings, constructions, graphs, dramatizations, and more on a daily basis. Review children's work to find evidence of what they know and can do.

> Sonal, Eli, Lucy, and Ashwin play a pattern game. As they play, they create a growing pattern. Eli brings his teacher over and shows that the pattern grows by one piece each time. The teacher photographs Eli with the growing pattern to document his learning.

Elicit Responses from Children: Responses can be answers to questions, parts of a discussion, participation in a task, and following directions. In children's responses, you can find clues to how they are thinking and what they understand.

After playing with Mix & Make Shapes™, the teacher asks Bella, Cameron, Blake, and Mia to help clean up. Each child is assigned a different shape to collect. Bella collects only the circles as asked, but the other three children pick up a random assortment of shapes. The teacher plans to continue work on shape recognition.

Keep a record of children's learning using each of these assessment methods. Write an observation on a sticky note, photograph a child's creation in the block area, and write children's responses to a group discussion. Use these records to assess a child's progress over time and to share information with parents and other educators.

Kindergarten Readiness

We have developed a Math Assessment, available online at getsetforschool.com/click which allows you to easily see which benchmarks your children have mastered. We include an assessment for general kindergarten readiness in our *My First School Book* activity book. It covers many of the important skills that children need to know for success in kindergarten. Children name pictures, colors, letters, and numbers. They also copy shapes, draw a person, and write their names. Send a copy of the completed assessment home with families at the end of the year.

Numbers & Math Pre-K Teaching Guidelines

Week	Monday	Tuesday	Wednesday
1	**Algebra** Classify Same or Different "One is Different, Can You Tell?" pp. 108–109	**Number & Operations** Demonstrate One-to-One Correspondence "Table's Ready!" pp. 34–35	**Geometry** Demonstrate In & Out "In or Out? Toss and Shout" pp. 74–75
2	**Number & Operations** Count a Set of Objects "Count With Me" pp. 36–37	**Geometry** Demonstrate In & Out "Cups & Caps" p. 75	**Number & Operations** Count a Set of Objects "Count & Chant" p. 37
3	**Algebra** Sort by Color "Moving Colors" p. 111	**Geometry** Demonstrate Before & After "Let's Make a Rainbow" pp. 76–77	**Number & Operations** Describe Cardinality "How Many in My Hat?" pp. 38–39
4	**Number & Operations** Count in Any Order "Count This Way & That" pp. 40–41	**Measurement** Make a Direct Comparison of Size "Compare & Share" pp. 126–127	**Number & Operations** Count in Any Order "Line Count" p. 41
5	**Geometry** Demonstrate Top, Middle & Bottom "Red Light, Green Light" pp. 78–79	**Algebra** Sort by Size "Fit Test" pp. 112–113	**Geometry** Demonstrate Top, Middle & Bottom "I Found the Top" p. 79
6	Discovery Play Tag Bags® p. 29	**Algebra** Sort by Size "Funnel Fun" p. 113	**Geometry** Demonstrate In & Out "Count It Out" p. 75
7	**Number & Operations** Make a Set "Match My Number" pp. 42–43	**Geometry** Demonstrate Above & Below, Over & Under "Listen & Pass" pp. 80–81	**Number & Operations** Make a Set "Set of Beads" p. 43
8	**Number & Operations** Recognize Quantities Without Counting "Hide & Peek" pp. 44–45	**Measurement** Compare Length Using Long & Short "Use Your Noodle" pp. 128–129	**Number & Operations** Use Ordinal Numbers "Tag Bag Line Up" pp. 46–47
9	**Algebra** Sort by Function "Fastener Sort" pp. 114–115	**Number & Operations** Compare Sets "More or Fewer?" pp. 48–49	**Geometry** Demonstrate Top, Middle & Bottom "What's in the Middle?" p. 79
10	**Data Representation & Probability** Move to Answer Questions "Eye Wonder" pp. 150–151	**Algebra** Describe a Simple Pattern "Repeat After Me" pp. 116–117	**Geometry** Demonstrate Before & After "What's on First?" p. 77

Thursday	Friday	Notes
Number & Operations Demonstrate One-to-One Correspondence "Socks & Shoes" p. 35	**Algebra** Classify Same or Different "More Money" p. 109	
Algebra Sort by Color "Color Sort" pp. 110–111	**Algebra** Classify Same or Different "Different Sizes" p. 109	
Geometry Demonstrate Before & After "Who's Before?" p. 77	**Number & Operations** Describe Cardinality "Tag Bag® Totals" p. 39	
Measurement Make a Direct Comparison of Size "Family Comparisons" p. 127	**Algebra** Sort by Color "Counting Colors" p. 111	
Algebra Sort by Size "Books in a Box" p. 113	**Number & Operations** Demonstrate One-to-One Correspondence "Counting Too" p. 35	
Measurement Make a Direct Comparison of Size "Comparing Shapes" p. 127	**Number & Operations** Count a Set of Objects "One in Each Box" p. 37	
Geometry Demonstrate Above & Below, Over & Under "Tower Power" p. 81	**Number & Operations** Describe Cardinality "Counting Buddies" p. 39	
Measurement Compare Length Using Long & Short "Longer & Shorter?" p. 129	**Number & Operations** Use Ordinal Numbers "Looking for Lines" p. 47	
Number & Operations Compare Sets "Count Each Set" p. 49	**Algebra** Sort by Function "Color Count" p. 115	
Algebra Describe a Simple Pattern "Make a Pattern" p. 117	**Geometry** Demonstrate Above & Below, Over & Under "Where Does It Go?" p. 81	

Week	Monday	Tuesday	Wednesday
11	**Data Representation & Probability** Move to Answer Questions "Movement Sort" p. 157	**Number & Operations** Identify Numerals "Name That Number" pp. 50–51	**Geometry** Left & Right "Shake Hands with Me" pp. 82–83
12	Discovery Play 1-2-3 Touch & Flip® Cards p. 33	**Measurement** Compare Length Using Long & Short "Check the Line" p. 129	**Number & Operations** Use Ordinal Numbers "Line Up Time" p. 47
13	**Data Representation & Probability** Move to Answer Questions "Square Dance" p. 151	**Geometry** Sort Shapes "Sides & Corners" pp. 84–85	**Algebra** Duplicate a Simple Pattern "Sounds Like This!" pp. 118–119
14	**Data Representation & Probability** Graph with Objects "Apples & Bananas" pp. 152–153	**Geometry** Match Shapes "Guess What?" pp. 86–87	**Number & Operations** Connect Numerals to Quantities "Show Me the Number!" pp. 52–53
15	**Data Representation & Probability** Graph with Objects "Compare by Pairs" p. 153	**Measurement** Compare Height Using Tall & Short "Tall Towers" pp. 130–131	**Number & Operations** Identify Numerals "Continue On" p. 51
16	**Data Representation & Probability** Graph with Objects "More to Eat" p. 153	**Geometry** Match Shapes of Different Sizes "Are You My Match?" pp. 88–89	**Number & Operations** Connect Numerals to Quantities "Play a Game" p. 53
17	**Number & Operations** Write Numerals "Wet, Dry, Try" pp. 54–55	**Geometry** Move Shapes to Match "Puzzle Moves" pp. 90–91	**Number & Operations/** ***I Know My Numbers* booklet 1** Write Numerals "*I Know My Numbers*" p. 55
18	Discovery Play 4 Squares More Squares® p. 71	**Geometry** Match Shapes "A Funny Face" p. 87	**Number & Operations** Recognize Quantities Without Counting "Tag Bag Flash" p. 45
19	*I Know My Numbers,* booklet 2	**Geometry** Describe Circles "Round & Round We Go" pp. 92–93	**Number & Operations** Label Sets "Write How Many" pp. 58–59
20	**Data Representation & Probability** Explore Pictographs "Dog or Fish?" pp. 154–155	**Measurement** Compare Weight Using Heavy & Light "Heavy or Light?" pp. 132–133	**Geometry** Describe Rectangles "Our Rectangle Book" pp. 94–95

Thursday	Friday
Number & Operations **Identify Numerals** "Counters in a Cup" p. 51	**Geometry** **Demonstrate Left & Right** "Learn Left" p. 83
Number & Operations **Count in Any Order** "Tag Bag Dot Count" p. 41	**Algebra** **Describe a Simple Pattern** "Bigger Pattern Units" p. 117
Number & Operations **Compare Sets** "Let's Add More" p. 49	**Algebra** **Duplicate a Simple Pattern** "Copy Cat" p. 119
Geometry **Match Shapes** "Match My Shape" p. 87	**Number & Operations** **Connect Numerals to Quantities** "Clap a Number" p. 53
Measurement **Compare Height Using Tall & Short** "Taller Than Our Plant" p. 131	**Number & Operations** **Make a Set** "Flip, Check, & Pass" p. 43
Geometry **Match Shapes of Different Sizes** "My Shapes Match" p. 89	**Measurement** **Compare Height Using Tall & Short** "Taller Towers" p.131
Geometry **Move Shapes to Match** "Match My Shape" p. 91	**Algebra** **Sort By Function** "Number Match" p. 115
Algebra **Duplicate a Simple Pattern** "Stringing a Pattern" p. 119	**Geometry** **Demonstrate Left & Right** "Left Hand, Right Hand" p. 83
Geometry **Describe Circles** "Curious Curves" p. 93	**Number & Operations** **Label Sets** "More Labels" p. 59
Measurement **Compare Weight Using Heavy & Light** "What's in Your Box?" p. 133	**Geometry** **Describe Rectangles** "Count How Many" p. 95

Notes

Week	Monday	Tuesday	Wednesday
21	*I Know My Numbers,* booklet 3	**Geometry** **Describe Squares** "1 Purple, 2 Purple, 3 Purple, 4" pp. 96–97	**Measurement** **Compare Width Using Narrow & Wide** "Let's Decide, Narrow or Wide?" pp. 134–135
22	**Data Representation & Probability** **Explore Pictographs** "Class Vote" p. 155	**Geometry** **Describe Triangles** "Terrific Triangles" pp. 98–99	**Measurement** **Compare Weight Using Heavy & Light** "What's in Your Bag?" p. 133
23	*I Know My Numbers,* booklet 4	**Number & Operations** **Combine Sets** "Count Them All" pp. 60–61	**Measurement** **Compare Capacity Using More & Less** "What Holds More?" pp. 136–137
24	Discovery Play Mix and Make Shapes™ p. 73	**Measurement** **Compare Capacity Using More & Less** "Test It!" p. 137	**Geometry** **Match Shapes of Different Size** "I Like Squares, Too" p. 89
25	*I Know My Numbers,* booklet 5	**Algebra** **Explore Growing Patterns** "Teacher, May I Grow a Pattern?" pp. 120–121	**Geometry** **Move Shapes to Match** "Puzzle Challenge" p. 91
26	**Data Representation & Probability** **Explore Pictographs** "Choice Challenge" p. 155	**Measurement** **Order By Size** "From Bears to Chairs" pp. 138–139	**Geometry** **Describe Rectangles** "Stop & Turn" p. 95
27	*I Know My Numbers,* booklet 6	**Geometry** **Recognize Shapes in a Group** "Simon Says Shapes" pp. 100–101	**Algebra** **Explore Growing Patterns** "Figure It Out" p. 121
28	**Number & Operations** **Take Objects Away** "How Many Are Left?" pp. 62–63	**Measurement** **Explore Area** "Cover & See" pp. 140–141	**Number & Operations** **Take Objects Away** "Dumplin' Song" p. 63
29	*I Know My Numbers,* booklet 7	**Measurement** **Use Nonstandard Units of Measurement** "Measuring with Bags" pp. 142–143	**Algebra** **Explore Patterns in the Real World** "Pattern Day" pp. 122–123
30	**Data Representation & Probability** **Identify Events as Likely or Unlikely** "Snow Boots or Sunglasses?" pp. 156–157	**Measurement** **Order By Size** "From Bears to Buttons" p. 139	**Number & Operations** **Take Objects Away** "Take Away More" p. 63

Thursday	Friday	Notes
Geometry Describe Squares "Building a Big Square" p. 97	**Measurement** Compare Width Using Narrow & Wide "Size Sort" p. 135	
Geometry Describe Triangles "Sing & Draw" p. 99	**Measurement** Compare Width Using Narrow & Wide "Word Time" p. 135	
Number & Operations Combine Sets "Place & Count" p. 61	**Measurement** Compare Capacity Using More & Less "Explore More" p. 137	
Number & Operations Combine Sets "Rowboat, Rowboat" p. 61	**Geometry** Sort Shapes "How Many Can You Find?" p. 85	
Algebra Explore Growing Patterns "Double Dip" p. 121	**Geometry** Describe Circles "Is It a Circle?" p. 93	
Measurement Order By Size "From Bears to Squares" p. 139	**Geometry** Describe Squares "Cover the Squares" p. 97	
Geometry Recognize Shapes in a Group "Body Shapes" p. 101	**Number & Operations** Label Sets "Clap & Trace" p. 59	
Measurement Explore Area "Different Ways" p. 141	**Geometry** Describe Triangles "I Can Make a Triangle" p. 99	
Measurement Use Nonstandard Units of Measurement "Standard Units" p. 143	**Algebra** Explore Patterns in the Real World "Pattern Hunt" p. 123	
Measurement Explore Area "How Many Squares?" p. 141	**Number & Operations** Write Numerals "Door Tracking" p. 55	

Week	Monday	Tuesday	Wednesday
31	*I Know My Numbers*, booklet 8	**Number & Operations** Share a Set Evenly "Everyone Gets the Same" pp. 64–65	**Geometry** Identify Shapes in Objects "Snap a Shape" pp. 102–103
32	**Data Representation & Probability** Identify Events as Likely Or Unlikely "Spinning to Predict" p. 157	**Geometry** Identify Shapes in Objects "My Book About Shapes" p. 103	**Measurement** Sequence Events "Step by Step" pp. 144–145
33	*I Know My Numbers*, booklet 9	**Number & Operations** Divide a Whole into 2 Halves "Half & Half" pp. 66–67	**Measurement** Connect Time & Events "Day & Night Charades" pp. 146–147
34	**Data Representation & Probability** Identify Events as Likely Or Unlikely "Certain or Impossible" p. 157	**Geometry** Explore 3-D Shapes "Stack It Up" pp. 104–105	**Measurement** Use Nonstandard Units of Measurement "Measure Volume" p. 143
35	**Number & Operations/** ***I Know My Numbers* booklet 10** Review Numerals "9, 10, A Big Fat Hen" pp. 56–57	**Geometry** Sort Shapes "Sort Some More!" p. 85	**Number & Operations** Review Numerals "Fun Practice" p. 57
36	**Measurement** Sequence Events "Line it Up"™ p. 145	**Number & Operations** Review Numerals "Without a Trace" p. 57	**Algebra** Explore Patterns in the Real World "Nature Patterns" p. 123

Thursday	Friday	Notes
Number & Operations Share a Set Evenly "Share Other Things" p. 65	**Geometry** Identify Shapes in Objects "Real World Shapes" p. 103	
Number & Operations Recognize Quantities Without Counting "Match My Dots" p. 45	**Measurement** Sequence Events "Mixed-Up Snack" p. 145	
Number & Operations Divide a Whole into 2 Halves "Cooking Cutting" p. 67	**Measurement** Connect Time & Events "Things We Do" p. 147	
Geometry Explore 3-D Shapes "Hunting for Shapes" p. 105	**Number & Operations** Share a Set Evenly "Share with More Friends" p. 65	
Geometry Recognize Shapes in a Group "More Simon Says" p. 101	**Measurement** Connect Time& Events "What Do You Do?" p. 147	
Geometry Explore 3-D Shapes "Shapes on My Shapes" p. 105	**Number & Operations** Divide a Whole into 2 Halves "A New Fold" p. 67	

Math Words for Children

These "Math Words for Children" are organized alphabetically. Use these words throughout multiple domains during activities and throughout your day.

A

above – higher up than, or over
after – later than; behind

B

before – sooner or earlier than; in front of
below – lower than; beneath
big/bigger – large in size
bottom – the lowest part of something

C

certain – definitely will happen
check – to look back at something to make sure it's correct
compare – to look at two things and see similarities or differences
corner – the point where two sides meet
count – to say numbers in order; to figure out how many there are of something
counting on – to continue to count from where you stopped
cover – to put something over
curve – a line that turns or bends but doesn't have corners

D

day – the time when the sun shines
different – not the same

F

fewer – smaller in number or amount
first – the beginning person or thing
fit – to be the right size and shape
flip – to turn over

G

growing – increasing in size, length, or amount

H

half – one of two equal parts
heavy – weighing a lot; not light
height – the measurement from top to bottom
holds less – contains a smaller amount
holds more – contains a larger amount
how many – asks you to find a total; what number or amount in all

I

impossible – definitely will not happen
in – within something; opposite of out
in all – all together; total amount

L

large – bigger than usual
last – the person or thing at the end; final
left – an amount remaining; the opposite of something on the right
length – distance from one end to the other; how long something is
light – having little weight; not heavy
likely – probably will happen
little – small in size or amount; not much
long/longer/longest – having length; a great distance

M

match – something that is the same as the other; to find things that are the same
medium – something that is in the middle, between big and little
middle – halfway between two objects, points, or times
more/most – greater in number or size when compared to another amount

N

narrow – small in width; not wide
next – coming just after or following
night – time when the sun doesn't shine, when it is dark
not enough – needing more

O

opposite – being on the other side of something
out – outside of something; opposite of in
over – above; on top of something

P

pattern – a repeating collection of colors, shapes, sounds, or figures

R

repeat – to say or do something again
right – on the side opposite the left
roll – to move around and around
round – shaped like a circle or ball; having a curved edge or surface
row – a line of things (horizontal)

S

same – exactly alike
second – ordinal number for the position after first
shape – the form of an object (circle, square, triangle)
share – to give a part to someone else
short/shorter/shortest – having little length or height; not long or tall
side – the edge of a shape or a surface of an object
size – how big or small something is
small/smaller/smallest – reduced in size; little
sort – to separate things by attribute or characteristic
stack – to place objects one on top of another
straight – having no curves, bends, or corners

T

take away – to subtract; to make a quantity smaller
tall/taller – having great height; not short
third – ordinal number for the position after second and before fourth
too many – having more than enough
top – the highest point or part of something
total – the whole amount; the sum
tower – a building or stack that is taller than it is long or wide
turn – to spin or to rotate

U

under – below or beneath

unlikely – probably will not happen

W

weight – how much something weighs

whole – all or total; all the parts

wide – large from side to side

width – the measurement from side to side

Glossary for Educators & Parents

A

AB/ABB – letters often used to define a pattern

algebra – math that uses letter symbols to represent numbers or patterns

area – the amount of space inside a shape

arithmetic – math that uses whole numbers, decimals, and fractions; it includes the operations of addition, subtraction, multiplication, and division

asymmetry – not having two sides the same: for example, numerals 2, 3, 4, 5, 6, 7, 9 are asymmetrical. Asymmetrical numbers, letters, or shapes may need to be flipped to fit in a puzzle

attribute – a characteristic or quality of an object

B

benchmark – A detailed description that outlines what a student is expected to know at a particular grade, age, or developmental level

C

cardinality – the counting principle that the last number said is the total number of items in a set

concrete objects – real things that children can physically touch (block, ball)

cooperative learning – an instructional method that encourages children to work and learn together

D

data representation – a way of displaying information visually (bar graphs, pictographs)

direct comparison – when children compare two or more objects or ideas

domain – a group of related learning skills

F

formative assessments – assessments that provide feedback to the teacher for the purpose of improving instruction

G

geometry – math that deals with area, positions, and shapes

I

independent learning – knowledge that children acquire by themselves without explicit teaching

informal assessments – casual observations or ways to collect information about children's learning

L

learning skills – a detailed description that outlines what a student is expected to know at a particular grade, age, or developmental level

M

manipulatives – objects such as blocks, counters, beads, etc. used to explore ideas and solve problems

N

nonstandard units – informal units of measure (e.g., child's hand, straws) not found in the customary or metric systems

numbers – a way to describe quantity or amount
- **cardinal numbers** (one, two, three) numbers used for counting objects in a set
- **counting numbers** (one, two, three) numbers used for counting objects in a set
- **ordinal numbers** (1st, 2nd, 3rd - first, second, third) numbers used for telling order in a sequence
- **numerals** (0, 1, 2, 3) number symbols (not words)

O

one-to-one correspondence – the matching of one thing to another (matching three counters to three circles)

operations – addition, subtraction, multiplication, and division

ordinal numbers – numbers used to tell a position in a sequence (1st, 2nd, 3rd,...)

orientation – the position or direction of an object or symbol faces

P

pattern – a sequence that repeats, such as of color, sound, movement, number, size, shape, or objects
- **AB/AABB:** letters often used to define a pattern. The letters can be substituted for any item (manipulatives, geometric shapes, colors, etc.)
- **repeating pattern:** a type of pattern that duplicates the starting pattern unit AB, AB, AB
- **extending patterns:** to continue a pattern AB, AB, AB...
- **growing pattern:** a type of pattern in which a new part is added to the previous section each time it restarts A, AB, ABC, ABCD, ABCDE...

pattern unit – the part of the pattern that repeats

pictograph – a chart that shows numerical information using picture symbols

prior knowledge – information and skills that children have from previous experiences

problem solving – using various methods to determine the best answer for a problem

R

reversal – a letter or number symbol that is facing the wrong direction

S

self-discovery – a time for children to discover things on their own about a topic

sorting – arranging items into groups by common attributes

spatial reasoning – the ability to interpret and make drawings or visualize movement of objects

standard units – customary and metric units of measure (rulers, inches, tons, grams, yardstick, centimeters, etc.)

symbol – a letter, number, or mark that represents something

symmetry – having two sides the same: for example, 0 and 8 are symmetrical numbers

T

transformations – changes in position that include slides, turns, and flips
- **slide:** moving in a direction without turning – as in sliding a circle puzzle piece into place
- **turn:** rotating, as in turning a symmetrical square puzzle piece to make it fit
- **flip:** turning over, as in flipping an asymmetrical L piece to make it fit

Z

zero – the number for nothing in a set

Index

References

Alliance for Childhood. 2006. "*A call to action on the education of young children.*" Retrieved from: www.allianceforchildhood.org.

Barnett, W.S., J.T. Hustedt, A.H. Friedman, J.S. Boyd, and P. Ainsworth. 2007. *The State of Preschool 2007.* New Brunswick, NJ: National Institute for Early Education Research.

Bergen, D. 2002. "The Role of Pretend Play in Children's Cognitive Development." *Early Childhood Research and Practice* 4(1): 2-15.

Berk, L.E., T.D. Mann, and A.T. Ogan. 2006. "Make-believe Play: Wellspring for the Development of Self-regulation." *Play = learning: How Play Motivates and Enhances Children's Cognitive and Social-Emotional Growth.* New York: Oxford University Press.

Boaler, J. 2008. *What's Math Got to Do With It?* New York: Penguin Group.

Boggan, M., S. Harper, and A. Whitmire. 2010. "Using Manipulatives to Teach Elementary Mathematics." *Educational Research* 3: 1-6. http://www.aabri.com/manuscripts/10451.pdf.

Bowman, B. T., S. M. Donovan, and S. M. Burns, eds. 2000. "Commission on Behavioral and Social Sciences and education and National Research Council." *Eager to Lean: Educating Our Preschoolers.* Washington D.C.: National Academy Press.

Boyd, J., S. W. Barnett, E. Bodrova, D. Leong, and D. Gomby. 2005. *Promoting Children's Social and Emotional Development through Preschool Education.* New Brunswick, NJ: National Institute for Early Education Research.

Brenneman, K., J. Stevenson-Boyd, and E.C. Frede. 2009. *Policy Brief – Math and Science in Preschool: Policies and Practice.* Issue 19. New Brunswick, NJ: National Institute for Early Education Research.

Burns, M. 2007. *About Teaching Mathematics: A K-8 Resource.* Sausalito, CA: Math Solutions.

Cain-Caston, M. 1996. Manipulative queen [electronic version]. *Journal of Instructional Psychology* 23(4): 270-274.

Clements, D.H., and M.T. Battista. 1992. "Geometry and Spatial Reasoning." *Handbook of Research on Mathematics Teaching and Learning.* NY: Macmillan.

Clements, D.H., & Sarama, J. 2000. "Young Children's Ideas about Geometric Shapes." *Teaching Children Mathematics* 6(8): 482.

Clements, D.H., Sarama, J., DiBiase, A. 2004. "Measurement in Pre-K to Grade 2 Mathematics." *Engaging Young Children in Mathematics: Standards for Early Childhood Mathematics Education.* Mahwah, NJ: Lawrence Erlbaum Associates, Inc.

Copley, J. 2000. *The Young Child and Mathematics.* Washington D.C.: National Association for the Education of Young Children.

Copple, C., and S. Bredekamp. 2009. Developmentally Appropriate Practice in the Preschool Years-Ages 3–5: Examples to Consider. Washington, D.C.: National Association for the Education of Young Children.

Coulter, D. 1995. Music and the making of the mind. *Early Childhood Connections: The Journal of Music- and Movement-Based Learning* 1: 22-26.

Cross, C. T., T. A. Woods, and H. Schweingruber, eds. 2009. *Mathematics Learning in Early Childhood: Paths Toward Excellence and Equity.* Washington, D.C.: The National Academies Press.

Duncan G.J., C.J. Dowsett, A. Claessens , K. Magnuson, and A.C. Huston. 2007. "School Readiness and Later Achievement." *Developmental Psychology* 43(6): 1428-1446.

Frede, E., and D.J. Ackerman. 2007. "Preschool Curriculum Decision-Making: Dimensions to Consider." *Preschool Policy Brief.* 3(12).

Geist, E. 2008. *Children are Born Mathematicians: Supporting Mathematical Development, Birth to Age 8.* NJ: Prentice Hall.

Gesell, A. 1940. *The First Five Years of Life: A Guide to the Study of the Preschool Child.* NY: Harper and Row.

Hannibal, M. A. 1999. "Young Children's Developing Understanding of Geometric Shapes." *Teaching Children Mathematics* 5(6): 353.

Healy. 2004. *Your Child's Growing Mind: Brain Development and Learning From Birth to Adolescence.* NY: Three Rivers Press.

Hirsh-Pasek, K., R. Golinkoff, L. Berk, and D. Singer. 2008. *A Manifesto for Playful Learning in Preschool: Presenting the Scientific Evidence.* NY: Oxford.

Howell, S., and C. Kemp. 2009. "A Participatory Approach to the Identification of Measures of Number Sense in Children Prior to School Entry." *International Journal of Early Years Education* 17(1): 47-65.

Isenberg, J., and N. Quisenberry. 2002. *Play: Essential for All Children. A position paper of the Association for Childhood Education International.* Retrieved from: www.acei.org/playpaper.htm.

Jensen, E. 2001. *Arts with the Brain in Mind.* Alexandria, VA: Association for Supervision and Curriculum Development.

Jones, E., and R. Cooper. 2006. *Playing to Get Smart.* NY: Teachers College Press.

Linder, S. M., and B. Powers-Costello. 2011. "Mathematics in Early Childhood: Research-Based Rationale and Practical Strategies." *Early Childhood Education Journal* 39: 29-37.

McAfee, O., D. Leong, and E. Bodrova. 2004. *Basics of Assessment: A Primer for Early Childhood Educators.* Washington, D.C.: National Association for the Education of Young Children.

National Association for the Education of Young Children and the National Council of Teachers of Mathematics. 2003. Joint position statement: *Early Childhood Mathematics: Promoting Good Beginnings.* Washington, DC.

National Association for the Education of Young Children. 2009. Position statement: *Developmentally Appropriate Practice in Early Childhood Programs Serving Children from Birth through Age 8.* Washington, DC.

National Association for the Education of Young Children and the National Association of Early Childhood Specialists in State Departments of Education. 2003. Joint position statement: *Early childhood curriculum, assessment, and program evaluation: Building an effective, accountable system in programs for children birth through age 8.*

National Association of State Boards of Education. 2006. *Fulfilling the Promise of Preschool: The Report of the NASBE Study Group on Creating High-Quality Early Learning Environments.* National Child Care Information and Technical Assistance Center.

National Council of Teachers of Mathematics. 2000. *Early Learning Standards in Mathematics.* Reston, VA: National Council of Teachers of Mathematics.

National Council of Teachers of Mathematics. 2006. Curriculum focal points for Pre-kindergarten through grade 8 mathematics. Reston, VA. Retrieved 6/14/2009 from the web at www.nctm.org/standards/focalpoints.aspx?id=300

National Research Council. 2009. *Mathematics learning in early childhood: paths toward excellence and equity.* Committee on Early Childhood Mathematics. Washington, DC: The National Academies Press.

Neuman, S. B., C. Carol, and S. Bredekamp. 2004. *Learning to Read and Write: Developmentally Appropriate Practices for Young Children.* Washington, D.C.: National Association for the Education of Young Children.

Newcombe, N.S., and A. Frick. 2010. "Early Education for Spatial Intelligence: Why, What, and How." *Mind, Brain, and Education* 4(3): 102-111.

Olsen, J. Z., and E. F. Knapton. 2008. *Pre-K Teacher's Guide.* Cabin John, MD: Handwriting Without Tears.

Pica, R. 2008. "Learning by Leaps and Bounds: Why Motor Skills Matter." *Young Children* 63(4): 48-9.

Platt, J.E., and S. Cohen. 1981. "Mental Rotation Task Performance as a Function of Age and Training." *Journal of Psychology: Interdisciplinary and Applied* 108: 173-178.

Rudd, L. C., M.C. Lambert, M. Satterwhite, and A. Zaier. 2008. "Mathematical Language in Early Childhood Settings: What Really Counts?" *Early Childhood Education Journal* 36: 75-80.

Sadler, F. H. 2009. "Help! They Still Don't Understand Counting." *TEACHING Exceptional Children Plus* 6(1): 3.

Seo, K.-H,. and H.P. Ginsburg. 2004. *Engaging Young Children in Mathematics Education.* Mahwah, NJ: Lawrence Erlbaum.

Singer, D.G., R. Golinkoff, and K. Hirsh-Pasek, Eds. 2006. *Play = Learning: How Play Motivates and Enhances Children's Cognitive and Social-Emotional Growth.* NY: Oxford University Press.

Singer, D. G., J.L. Singer, S.L. Plaskon, and A.E. Schweder. 2003. *All Work and No Play: How Educational Reforms are Harming our Preschoolers.* Westport, CT: Praeger.

Stevenson, H.W., and R.S. Newman. 1986. "Long-term Prediction of Achievement and Attitude in Mathematics and Reading." *Child Development 57*: 646-659.

Taylor-Cox, J. 2003. "Algebra in the Early Years? Yes!" *Young Children* 58(1), 14-21.

UVA Today. 2009. U. Va. "Researchers Receive $1M to Study Connection Between Fine Motor and Math Skills." Retrieved from: www.virginia.edu/uvatoday/newsRelease.php?id=10150

Van De Walle, J. A. 2004. *Elementary and Middle School Mathematics Teaching Developmentally.* Boston: Pearson.

Wardle, F. 2007. "Math in early childhood." *Exchange,* March/April. Retrieved from: https://secure.ccie.com/library/5017455.pdf.

West, J., K. Denton, and E. Germino-Hausken. 2000. *America's kindergarteners. Statistical analysis report.* Washington, D.C.: National Center for Education Statistics.

West, J., K. Denton, and L. Reaney. 2000. *The Kindergarten Year.* Washington, D.C.: U.S. Department of Education.

Wright, T. S., and S.B. Neuman. 2009. *Preschool Curriculum: What's In It for Children and Teachers,* by Peg Griffin and Catherine King. Washington, D.C.: The Albert Shanker Institute.